Per Axel Rydberg

Flora of the Black Hills of South Dakota

Per Axel Rydberg

Flora of the Black Hills of South Dakota

ISBN/EAN: 9783337268473

Printed in Europe, USA, Canada, Australia, Japan

Cover: Foto ©berggeist007 / pixelio.de

More available books at **www.hansebooks.com**

ISSUED JUNE 13, 1896.

FLORA OF THE BLACK HILLS OF SOUTH DAKOTA.

BY

P. A. RYDBERG.

WASHINGTON:
GOVERNMENT PRINTING OFFICE.
1896.

LETTER OF TRANSMITTAL.

U. S. DEPARTMENT OF AGRICULTURE,
DIVISION OF BOTANY,
Washington, D. C., February 14, 1896.

SIR: I have the honor to transmit herewith, for publication as Contributions from the United States National Herbarium, Volume III, No. 8, a report entitled "Flora of the Black Hills of South Dakota," by Mr. P. A. Rydberg. This report is based upon a botanical exploration made in that region by Mr. Rydberg in 1892, as a field agent of the Department of Agriculture.

Respectfully, FREDERICK V. COVILLE,
Botanist.

Hon. J. STERLING MORTON,
Secretary of Agriculture.

CONTENTS.

ILLUSTRATIONS.

V

FLORA OF THE BLACK HILLS OF SOUTH DAKOTA.

By P. A. Rydberg.

ITINERARY.

On May 26, at noon, I left Lincoln by the Burlington and Missouri River Railroad to spend three months in the Black Hills of South Dakota. A few days before, I had received a commission from the United States Department of Agriculture as an agent of the Division of Botany for the purpose of making an investigation of the flora of the Black Hills. Up to July 3, I was accompanied by Mr. A. E. Wagner, a graduate student of the University of Nebraska. In the morning of May 27 we arrived at Edgemont, where the railroad crosses the Cheyenne River. I availed myself of frequent opportunities to collect along the railroad track. Arriving at Custer, which point was taken as the base of operations, being about the center of the Hills, we remained there until June 7, collecting in all directions from 4 to 6 miles from town; but the number of plants in flower was not large. On July 7 we moved our camp to an artificial body of water (Sylvan Lake) in the mountains not far from Harneys Peak. On June 10 we returned to Custer and took the train for Hot Springs, where we stayed, collecting, till June 21. Then taking the Fremont, Elkhorn, and Missouri Valley Railroad we followed the eastern foothills up to Elk River, stopping on the way at the following places: Buffalo Gap, June 21; Hermosa, June 22 to 24; Rapid City, June 25 to 26; and Piedmont, June 27 to 29.

On the morning of June 29 we bought tickets on the Black Hills and Fort Pierre Railroad to Runkels, but stepped off the train at Jones, making our way up the remaining length of the famous Elk Canyon on foot. We remained at Runkels through the next day, removing to Lead City the morning of July 1, at which point my camp remained till July 9. Mr. Wagner set out on his return to Lincoln July 3. From this base I visited the region around Deadwood, July 5; Whitewood, July 7; and Terrys Peak, July 8. On the 9th of July, I took the Burlington and Missouri River Railroad south, stopping at Rochford July 9 to 13, and at Custer July 13 to August 1. From the latter point I visited Oreville on July 17; Sylvan Lake and Harneys Peak, July 18 to 21; made a drive of 22 miles down French Creek July 22 and 23, and one into the Limestone District near the Wyoming line, July 25 to 30, extending probably 30 to 35 miles northwest of Custer.

On August 1, I moved my camp to Minnekahta, visiting Pringle on the 6th. I then joined a party of naturalists from the University of Nebraska, in whose company I remained for the rest of the season. We camped at Hot Springs till August 11, when we moved to Custer. From this point, after visiting the Harneys Peak region on August 17 to 18, we returned to Lincoln on August 22.

GEOGRAPHY.

The Black Hills are on the boundary line between South Dakota and Wyoming, the larger part lying within the former State. The center is a little east of the intersection of the forty-fourth parallel and one hundred and fourth meridian. The Black Hills constitute an isolated range about 120 miles long north-northwest and south-southeast, and 40 to 50 miles wide east and west. A little northwest of the Black Hills and separated from them only by the narrow valley of the Belle Fourche, is another much smaller spur, the Little Missouri Mountains, evidently belonging to the same range. The nearest mountains except those mentioned are the Big Horn Mountains to the west and the Laramie Mountains to the southwest. These are at a distance of 150 to 200 miles and separated from the Black Hills, the former by the valleys of the Little Missouri and Powder rivers, the latter by those of the Cheyenne and North Platte. There are no mountains to the north, east, or south.

Not only are the Black Hills an isolated range, but the surrounding high table-land is deeply cut on all sides by the branches of the Cheyenne River. The head of Inyankara Creek is due west of the center of the Hills. The creek runs in a northwesterly direction till it empties into the Belle Fourche. This runs northeast and then southeast, emptying into Cheyenne River. The bend is north of the range. Not far from the head of the Inyankara are the springs of Beaver Creek, a stream which flows south into Cheyenne River. The latter runs south of the Hills, then changes its course to northeast till it joins the Belle Fourche, and finally empties into the Missouri River.

GEOLOGY.

In order to compare the geological and floral districts of the Black Hills I give a summary of the geology of the region derived from the report by Henry Newton on the Geology and Resources of the Black Hills of Dakota.[1]

---

[1] U. S. Geog. and Geol. Survey of the Rocky Mountain Region, 1880.

Mr. Newton gives a synopsis of the strata of the Black Hills, of which the following is a short abstract:

Ages.	Feet.	Strata.
Cenozoic:		
Miocene	200	Clays and conglomerates.
Cretaceous	900	Clays and shales; sandstone.
Mesozoic:		
Jura	200	Clays and marls, with some limestone.
Red Beds	340	Red clay.
Carboniferous	690	Sandstones and limestones
Paleozoic:		
Silurian (Potsdam)	250	Silicious sandstones and conglomerates.
Archæan (?)		Slates and schists, with intrusive granite.

Of these the Mesozoic and Paleozoic strata are resting conformably upon each other, the Cenozoic and Archæan are not. The Black Hills have been formed by the uprising of the Archæan rocks, which lifted up and broke through the overlying strata. That this uprising must have taken place after the Cretaceous and before the Miocene formations can be seen from the fact that the strata of the latter do not conform with those of the former.

The Black Hills have received their present form by erosion. The softer rocks have worn away faster, leaving the harder standing out as ridges or crags. As the center is raised, the dip of the strata is outward and the outcroppings form concentric ovals. The Jurassic clays and marls, the clays of the Red Beds and the slates and schists of the Archæan formation are comparatively soft, while the Cretaceous sandstone and the granite spurs of Archæan formation are hard. The Miocene formation belongs to the surrounding plains and does not enter the Hills anywhere.

The outcrop of the Cretaceous sandstone forms the foothills, and that of the Jura formation and the Red Beds makes the so-called "Race Track," a more or less continuous valley between the foothills and the hills proper. The Race Track is much broader on the south side and forms the larger part of the "Minnekahta Plains." On the east side the Carboniferous limestone belt forms a series of hills, and on the west side, where the strata are nearly horizontal, a broad plateau, the "Limestone District." As the slates and schist are comparatively soft, the center of the Black Hills is lower, except where the granite or igneous rocks come to the surface and form the highest peaks in the range.

ALTITUDES.

The plains at the base of the Black Hills have an altitude of about 1,000 meters. The highest point is the top of Harneys Peak, one of the series of granite crags at about the center of the Hills. Its altitude is given differently. The most reliable measurements are without doubt those given by the United States Geological Survey, one of which makes it 7,368 feet, the other 7,440 feet, or, respectively, 2,245 and 2,267 meters. The people in the neighborhood insist upon calling it 8,000 feet. Next in height comes Crooks Tower, the highest point on the Limestone Plateau on the west side, with the altitude of 7,325 feet, or 2,232 meters. The altitude of Terrys Peak, one of the igneous cones in the Northern Hills, is 7,215 feet, or 2,200 meters, and Custers Peak, another igneous cone in that neighborhood, is slightly lower. The general height of the Black Hills is between 1,500 and 2,000 meters.

PRECIPITATION AND TEMPERATURE.

The precipitation of the Black Hills is very large compared with that of the surrounding country. There are no records of the annual rainfall in that part in which, judging from the nature of the flora and the luxuriant growth of the vegetation, the precipitation must be the largest, viz, in the Harney Mountain Range. In fact, there are no records from any place in the higher part of the Black Hills.

Table showing the average monthly and annual precipitation at the Weather Bureau stations in the Black Hills, compared with that of four stations on the plains and three stations within the Rocky Mountain region.

Station.	Altitude in meters	Year.	Jan.	Feb.	Mar.	Apr.	May	June	July	Aug	Sept
Deadwood, S. Dak	¹1,411	1878–87	1.25	1.21	1.99	5.17	4.65	3.73	2.61	2.23	1.06
Spearfish, S. Dak	1,140	1889–91	1.22	1.93	1.22	2.67	3.06	5.18	2.86	1.44	.89
Rapid City, S. Dak	973	{1881–84 1888–91}	.45	.83	1.16	2.03	4.30	3.91	2.14	1.59	.74
Fort Meade, S. Dak	²1,057	1879–91	.73	.63	1.07	2.38	4.02	3.22	2.40	1.96	.56
Fort Robinson, Nebr	1,150	1883–91	.50	.67	1.24	1.65	2.93	2.61	2.04	1.94	.53
Sidney, Nebr	1,248	{1872–80 1886–91}	.52	.47	.93	1.37	2.83	1.98	2.55	2.06	1.33
Fort Fetterman, Wyo	1,512	1868–82	.46	.67	1.19	1.68	2.51	1.26	1.59	.96	1.11
Fort Keogh, Mont	721	1881–90	.72	.53	.46	.81	2.10	2.75	.83	1.19	.79
Helena, Mont	¹1,399	{1866–69 1880–87}	1.73	.63	.70	1.57	1.50	2.47	.97	.74	1.37
Georgetown, Colo	2,586	1886–90	.42	.39	.74	1.29	1.75	.78	2.08	1.72	.79
Colorado Springs, Colo	1,821	1871–90	.20	.26	.46	1.72	2.26	1.69	3.45	2.14	1.28

Station.	Altitude in meters.	Year.	Oct.	Nov.	Dec.	Annual.	Greatest during any year. Year.	Greatest during any year. Inches.	Least during any year. Year.	Least during any year. Inches.
Deadwood, S. Dak	¹1,411	1878–87	1.58	1.26	1.51	28.48	1882	33.83	1880	19.20
Spearfish, S. Dak	1,140	1889–91	1.44	.53	1.14	22.68	1891	26.32	1890	20.73
Rapid City, S. Dak	973	{1881–84 1888–91}	.51	.35	.45	18.16	1888	22.75	1890	14.02
Fort Meade, S. Dak	²1,057	1879–91	.68	.48	.46	18.59	1883	27.05	1885	13.25
Fort Robinson, Nebr	1,150	1883–91	1.50	.57	.71	16.29	1887	25.25	1886	11.08
Sidney, Nebr	1,248	{1872–80 1886–91}	.78	.39	.24	14.55	1879	26.92	1880	7.61
Fort Fetterman, Wyo	1,512	1868–82	1.00	.80	.86	14.13	1871	17.86	1876	9.45
Fort Keogh, Mont	721	1881–90	.72	.37	.25	11.52	1891	14.18	1883	9.07
Helena, Mont	¹1,399	{1866–69 1880–87}	1.13	1.17	1.17	15.95	1866	22.11	1882	10.32
Georgetown, Colo	2,586	1886–90	.89	.78	.57	12.11	1891	16.42	1890	11.72
Colorado Springs, Colo	1,821	1871–90	.56	.37	.31	14.72	1872	18.56	1888	9.12

¹ The altitudes of Deadwood and Helena are those of the signal stations at those places. For the other places the altitudes of the railroad tracks at the stations are given.
² The altitude given for Fort Meade is that of Sturgis, the nearest point whose altitude is accessible.

The above table, which is an abstract from the records of the United States Weather Bureau, shows the average monthly and annual precipitation, given in inches, for four places in the region. Of these, Rapid City and Fort Meade are just outside the foothills. Spearfish in a canyon in the first range, and only Deadwood within the Hills proper; but the last is neither at a very great altitude nor even near the part that has the greatest precipitation. The table shows an increase of about 2 inches in the average annual precipitation for every 100 meters in altitude. If the precipitation of the higher parts of the Black Hills should be calculated on that basis, which would obviously be incorrect since many other meteorological conditions must be taken into consideration, the annual rain and snow fall around Custer would be about 32 inches and that of the Harney Mountains about 40 inches or more, and still I do not think these figures are overestimated. The signs are unmistakable that the precipitation of the region around Harneys Peak is much larger than that at Deadwood. The more luxuriant growth of the vegetation, notwithstanding the higher altitude, the great abundance of plants that need a humid climate, as for instance ferns, mosses, liverworts, and lichens, and the innumerable streams that originate there, show that the precipitation of the southern Hills must be greater than that of the northern.

For comparison I have included in the table the precipitation of four stations situated on the table-land between the Black Hills and the Rockies, and also of three stations within the Rocky Mountain region. It can be seen at a glance that the comparison is favorable to the Black Hills. It would be naturally anticipated that the precipitation would be greater here than on the nearly treeless plains, but that it should be so much greater than in the places cited in the Rockies is more unexpected. It is intimated that the rains from the Gulf of Mexico do not reach the Rocky Mountains, and that the rains in that region come from the west. A place situated on the east side of the mountain, as is the case with the three stations given, would receive very little rain, as the moisture would be combed out by the mountains.

Whatever the cause, the precipitation of the Black Hills is greater than that of certain places in the Rockies, as can also be seen from the table. The situation of Colorado Springs is similar to that of Rapid City or Fort Meade, and that of Georgetown or Helena can be compared with that of Deadwood. Even on the top of Pikes Peak, which takes away the rain from Colorado Springs, the precipitation is not greater than that of Deadwood, and I am sure that it would stand low in comparison with that of the Harney Mountains, if any records had been made at the latter place.

Table showing the average monthly maximum, mean, and minimum temperatures at the Weather Bureau stations in the Black Hills and the mean temperature at four stations on the surrounding plains.

Station.	Year.	Temperature.	Jan.	Feb.	Mar.	Apr.	May.	June.	July.	Aug.	Sept.
Deadwood, S. Dak. (altitude, 1,411 meters).	1878-87	Maximum	52	57	64	70	77	89	91	91	82
		Mean	21	23	32	40	50	60	65	65	54
		Minimum	17	16	3	16	27	39	44	41	31
Rapid City, S Dak. (altitude, 973 meters).	1881-83 / 1888-91	Maximum	57	60	69	-2	81	95	99	109	94
		Mean	20	22	33	46	53	64	71	71	61
		Minimum	18	-23	-2	13	27	41	46	45	34
Fort Meade, S Dak. (altitude of Sturgis, 1,057 meters).	1879-91	Maximum	53	62	68	80	85	94	98	100	91
		Mean	18	21	31	44	51	65	71	70	59
		Minimum	-22	-25	-6	15	27	39	45	41	51
Fort Robinson, Nebr. (altitude, 1,150 meters).	1883-91	Mean	21	24	34	48	57	67	73	70	61
Sidney, Nebr. (altitude, 1,248 meters).	1872-80 / 1886-91	Mean	● 21	26	32	43	53	66	72	70	59
Fort Fetterman, Wyo. (altitude, 1,512 meters).	1868-82 / 1890-92	Mean	24	28	33	43	55	67	74	70	56
Fort Keogh, Mont. (altitude 721 meters).	1878-92	Mean	11	16	32	47	56	67	74	72	60

Station.	Year.	Temperature.	Oct.	Nov.	Dec.	Average annual mean.	Highest maximum and lowest minimum. Degrees.	Highest maximum and lowest minimum. Date.	Average range.	Greatest range.
Deadwood, S. Dak. (altitude, 1,411 meters).	1878-87	Maximum	74	62	54	......	102	July, 1881	108	134
		Mean	44	33	23	42				
		Minimum	18	-1	13	......	-32	Feb., 1883		
Rapid City, S. Dak. (altitude, 973 meters).	1881-83 / 1888-91	Maximum	78	71	64	......	106	July, 1881	123	146
		Mean	48	34	31	46				
		Minimum	22	3	5	......	-40	Feb., 1883		
Fort Meade, S.Dak. (altitude of Sturgis, 1,057 meters).	1879-91	Maximum	80	68	61	......	108	July, 1881	125	145
		Mean	48	33	24	45				
		Minimum	16	5	13	......	-37	Dec., 1887		
Fort Robinson, Nebr. (altitude, 1,150 meters).	1883-91	Mean	49	36	29	47				
Sidney, Nebr. (altitude, 1,248 meters).	1872-80 / 1886-91	Mean	47		24	46				
Fort Fetterman, Wyo. (altitude, 1,512 meters).	1868-82 / 1890-92	Mean	45	33	25	46				
Fort Keogh, Mont. (altitude 721 meters).	1878-92	Mean	46	31	19	41				

From the accompanying table, also an abstract from the Weather Bureau records, but with the fractions omitted, it can be seen that the average temperature of the stations at the base is about the same as that of the surrounding plains. At Deadwood it is a few degrees lower, a decrease of about 1 degree for each 100 meters increase of altitude, and very likely in the higher hills the average temperature is much lower. The maximum heat is reached in August, the average being 100° F. in the foothills and 91° F. at Deadwood, but sometimes reaching higher, as seen in the table. The minimum is reached in February or January, being —25 and —23 degrees in the foothills, but only —17 degrees at Deadwood, probably on account of the more protected locality. The average difference between the maximum and minimum in a year or the range of variation is given in the tables; so also the difference between the highest maximum and the lowest minimum reached during the whole period of observation.

FLORAL DISTRICTS.

Relatively to the differences in topographical and climatic conditions and in vegetation, the Black Hills may be divided into five districts: Foothills, Minnekahta Plains, Harney Mountain Range, Limestone District, Northern Hills.

These districts do not coincide with the outcroppings of the different geological formations. They receive, however, their most prominent physical features from that formation which is best represented within the region, as for instance, the Minnekahta Plains from the Red Beds, the Limestone District from the Limestone Plateau, the Harney Range from the granite crags.

FOOTHILLS [1] AND SURROUNDING PLAINS.

The foothills are capped by the comparatively hard cretaceous sandstone. The plains outside the foothills are mostly covered by the overlying Miocene conglomerates and clays. In the canyons, along the water courses, and in other depressions the underlying thin Jurassic strata of clays and marls and the Red Beds are exposed. Although the foothills constitute the outcropping of an older formation their flora is essentially the same as that of the surrounding table-lands, which extend as valleys far in among the hills. The flora depends more on meteorological conditions than on the geological formation. As shown above, the annual rainfall at Rapid City, which is among the foothills, is much less than that of the Black Hills proper. In fact the conditions are much the same as in western Nebraska and eastern Wyoming. It is a dry region, with most of the rain falling in the spring, and a season of drought in July and August. A majority of the plants peculiar to the high, dry plains of Nebraska, Wyoming, and neighboring States were also found here. Most of these plants are endowed with characters that in one way or another reduce the evaporation to a minimum. These characteristic plants may be divided into the following groups:

(1) Very hairy plants, in many cases covered with a thick pannose pubescence. Such are:

Eriogonum flavum.
Eriogonum annuum.
Eriogonum multiceps.
Eriogonum pauciflorum.
Astragalus gilviflorus.
Eurotia lanata.
Plantago purshii.

Senecio canus.
Senecio plattensis.
Ecoteulus unttallianus.
Filago prolifera.
Spiraia lambertii sericea.
Artemisia frigida.

[1] The western foothills are in Wyoming. The work was confined to South Dakota, and hence this includes only the eastern foothills.

(2) Plants with a glaucous foliage having a hard epidermis:

Agropyron repens glaucum.
Elymus canadensis glaucifolius.
Yucca glauca.
Zygadenus venenosus.

Rumex venosus.
Adorium tenuifolium.
Argemone alba.
Viola nuttallii.

(3) Plants with white, often shreddy, stems:

Œnothera pallida.
Œnothera albicaulis.
Mentzelia decapetala.

Mentzelia nuda.
Mentzelia oligosperma.

(4) Plants in which the surface is reduced to a minimum, either by a special habit, as in the species of Opuntia and Cactus, or by the leaves being narrow and involute, as in the following:

Calamovilfa longifolia.
Lygodesmia juncea.

Carex filifolia.
Carex stenophylla.

(5) Plants with a deep-seated, enlarged root:

Ipomœa leptophylla.[1]

Psoralea esculenta.

Many of the plants belonging to the Dry-plain flora, and supposed to be of more southern or western range, extend into the foothills. Among these may be mentioned:

Jacksonia trachysperma.
Astragalus plattensis.
Astragalus racemosus.
Astragalus spatulatus.
Astragalus gracilis.
Astragalus microlobus.
Chenopodium fremonti incanum.

Psoralea cuspidata.
Adorium tenuifolium.
Peucedanum villosum.
Erigeron canus.
Erigeron flagellaris.
Croton texensis.
Sedum stenopetalum.

In many places in western Nebraska and South Dakota and eastern Wyoming there are no visible streams. The superfluous water is drained off by means of "sand draws." A sand draw is a subterranean stream. On the surface is seen only a broader or narrower band of pure sand, marking the channel. The water may sometimes be running 5 meters below the surface. Sand draws are found here or there among the foothills, but their place is mainly taken by the numerous streams running down from the hills. Many of these streams sink, however, and become sand draws before they reach Cheyenne River. Many plants either from the Black Hills proper or from the Missouri Valley have spread along the water courses. Among those which have ascended the streams may be mentioned:[2]

Ranunculus macounii.
Roripa nasturtium.
Œnothera sinuata.

Pentstemon grandiflorus.
Prunella vulgaris.
Polygonum lapathifolium.

[1] At the State University of Nebraska there is a root preserved which, in its dry state, is three-fourths of a meter long and 3.5 decimeters in diameter.
[2] Plants ranging across the continent or found as well in the Mississippi Valley as in the Rocky Mountains are mostly omitted.

Besides these most of the woody vegetation of the region, as:

Vitis vulpina, frost grape.
Populus deltoides, cottonwood.
Ulmus americana, elm.
Prunus virginiana, chokecherry.
Fraxinus pennsylvanica lanceolata, green ash.
Quercus macrocarpa, bur oak.
Salix fluriatilis, willow.
Ostrya virginiana, ironwood.
Cratægus macrantha, hawthorn.

Parthenocissus quinquefolia, Virginia creeper.
Acer negundo, box elder.
Prunus americana, plum.
Fraxinus pennsylvanica, red ash.
Salix cordata, willow.
Rosa woodsii, rose.
Celtis occidentalis, hackberry.
Celastrus scandens, woody bittersweet.

The last two in each column are local, having been observed only at one station each, viz. at Rapid City, Hot Springs, Hermosa, and Piedmont, respectively. The following have descended from the Black Hills proper.

Growing in canyons:

Populus tremuloides, quaking aspen.
Cornus baileyi, dogwood.
Berberis aquifolium, Oregon grape.

Salix bebbiana, willow.
Populus angustifolia, black cottonwood.

On the hills:

Pinus ponderosa scopulorum, Rocky Mountain yellow pine.

The following woody plants may be regarded as belonging to the region itself, that is, to the flora of the high plains and foothills:

Juniperus virginiana, red cedar.
Prunus demissa, western chokecherry.
Rhus trilobata, skunk brush.

Ribes cereum, squaw currant.
Ribes aureum, buffalo currant.
Rosa arkansana, prairie rose.

As objects of peculiar interest seen in this region may be mentioned a shrub of skunk brush, which had stems 3 meters high and one decimeter in diameter, a cottonwood that measured over 5.5 meters in circumference, 1.5 meters above ground, and had a branch below that height nearly 2 meters around, and another cottonwood on which all the leaves had a cuneate base. *Populus acuminata* was also rediscovered near Hot Springs.

Differently from the Black Hills proper the foothills are not covered by forest. Some of them are crowned by scattered pines mixed with a few cedars. The hills as well as the valleys are generally covered with grass. The principal grasses are:

Bouteloua oligostachya.
Bouteloua hirsuta.
Koeleria cristata.
Calamovilfa longifolia.

Bulbilis dactyloides.
Carex filifolia.
Andropogon scoparius.

With the exception of the last in each column they furnish good winter as well as summer pasture. The cattle and horses generally "range" the year around, and are often not given any hay except during snowstorms or other bad weather. The first four plants mentioned become self-cured during the dry season and are as good as hay. On account of the small rainfall and the season of drought, farming, as a rule, is not paying and the settlers have been forced to rely on stock raising. The little hay needed is cut along the streams. The principal hay grasses are:

Panicum virgatum.	Andropogon provincialis.
Agropyrum repens glaucum.	Phalaris arundinacea.
Elymus canadensis.	Calamagrostis canadensis.
Elymus virginicus.	Calamagrostis neglecta.
Poa nemoralis.	Panicularia nervata.

At Hot Springs a new Poa was found, described in this report under the name of *Poa pseudopratensis*, and here also *Savastana odorata* occurred.

MINNEKAHTA PLAINS.

The Minnekahta Plains are not plains such as we find in central Nebraska, but a high, rolling table-land, between the foothills to the south and the Harney Range to the north. Geologically they are made up of two formations. The southern part is an expansion of the so-called "Race Track" produced by the outcropping of the Red Beds, which is here wider than in any other part of the Hills.[1] As the vegetation nowhere fully covers the ground the whole landscape receives a peculiar reddish color. In the northern part the underlying carboniferous limestone comes to the surface. As the strata are lying comparatively undisturbed in their natural relation, the surface is less rugged than in other parts of the Black Hills, and there is here little difference in surface condition between the limestone formation and the Red Beds south of it, except in the color of the soil.

The Minnekahta Plains are crossed by the Burlington and Missouri River Railroad from a few miles south of the Minnekahta station to Pringle, where the road enters the mountain range. The plains are covered with grass and are mostly used as pastures, but part is under cultivation. The region seems to suffer somewhat from drought. I collected there in August, but found very little of interest. Woody plants were scarce. On the hills grew some pines, dwarf sumacs or skunk brush, and sand cherries; in the draws some box elders, cottonwoods, gooseberries, and plums. Among herbs of interest there occurred two stragglers from the South, viz: *Asclepias verticillata pumila*, and *Acerates auriculata*, and the following were abundant on the railroad embankment:

Amaranthus blitoides.	Solanum triflorum.
Setaria viridis.	Saponaria vaccaria.

Beside the common upland grasses, a few of special interest were collected, viz: *Poa feudleriana*, *Sporobolus heterolepis*, *Danthonia spicata*. The first is of a more western range and the others are from the East. All three were found in the neighborhood of Pringle. Otherwise the flora was much the same as in the foothills.

[1] On the east side it is narrow and its flora does not differ from that of the foothills.

473

The only really mountainous part of the Black Hills is between Pringle and Hill City, on the Burlington and Missouri River Railroad. Especially is the Harney Range, between Custer and the latter place, of a truly mountainous character. This district is a series of high, naked cliffs and crags, rising from 500 to 1,000 meters over the valleys, intermixed with smaller hills. The looser slates and schists of the Archaean age have worn and washed away, leaving the harder granite rocks standing out as gigantic prongs of the most fantastic shapes. In many cases the streams have hollowed out deep ravines and gulches. Where the granite rocks are less common broad valleys have been formed, which are often called "parks." The most important are Custer Park around the upper part of French Creek, Dodge Park around the heads of Red Canyon Creek, and Elk Prairie on the Upper Spring Creek.

The hills and the sides of the mountains are covered with woods, the valleys are open, rich grass lands, here and there under cultivation. The principal tree is the Rocky Mountain yellow pine (*Pinus ponderosa scopulorum*), the only tree that grows abundantly enough to make a forest. Lumbermen distinguish two varieties, in which I could see only individual variation. On the north side of the mountains, and even on the south side of the Harney Mountains at an elevation of about 900 meters above the level of French Creek and between 540 and 580 meters above the sea, there is also found spruce, but not, as one would expect, any of the Rocky Mountain species. It is the northern white spruce (*Picea canadensis*). But how did it come to the Hills? The pines have probably come from the west, from the Rockies, over the Big Horn or the Laramie mountains, and the hills of Wyoming. The deciduous trees have crept up the tributaries of the Cheyenne River. The spruce, which grows only in the highest part of the Hills could not have done either. The nearest point in the Rockies from which I have seen the white spruce reported is about 100 miles farther north and 400 or 500 miles farther west, viz, in the valley of Blackfoot River in western Montana. There are no high mountains north of the Black Hills, and the spruce apparently is not found growing anywhere else in the Dakotas or eastern Montana. Neither does it grow in the two mountain ranges named above nor in the Yellowstone National Park. It must have come to the Black Hills in prehistoric times, when Dakota had a colder climate and the woods extended over the plains, or else seeds must have been brought there by migratory birds. The juniper, a nearly prostrate form of *Juniperus communis*, is common on the knolls, but the red cedar *J. virginiana* is very rare. I saw only two stunted shrubs on the Buckhorn Mountains near Custer.

Of the deciduous trees there are:

Betula papyracea, canoe birch.
Betula occidentalis, western black birch.
Populus tremuloides, quaking aspen.

Salix bebbiana, willow.
Salix discolor, willow.
Salix cordata, willow.

13144—No. 8——2

Farther down along Squaw Creek occurred:

Quercus macrocarpa, bur oak.　　　　　　*Ulmus americana,* white elm.

Among shrubby plants may be mentioned:

Cornus stolonifera, dogwood.　　　　　　*Amelanchier alnifolia,* juneberry.
Ribes setosum, gooseberry.　　　　　　　*Corylus rostrata,* hazel.
Ribes oxycanthoides, gooseberry.　　　　*Opulaster opulifolius,* nine-bark.
Ribes cereum, squaw currant.　　　　　　*Opulaster monogynus,* nine-bark.
Ribes lacustre, swamp currant.
Shepherdia canadensis, Canadian
　Shepherdia.

The known range of the following Rocky Mountain plants is extended
by their discovery in the Black Hills on this trip:

Actæa spicata arguta.　　　　　　　　　*Aster sibiricus.*
Viola canina adunca.　　　　　　　　　　*Arnica alpina.*
Epilobium paniculatum.　　　　　　　　*Pyrola rotundifolia bracteata.*
Epilobium drummondii.　　　　　　　　　*Myosotis sylvatica.*
Dodecatheon pauciflorum.　　　　　　　*Wulfenia rubra.*
Aconitum fischeri.　　　　　　　　　　　*Astragalus aboriginum glabriusculus.*
Leucocrinum montanum.　　　　　　　　*Helianthemum majus.*
Arenaria verna hirta.

Of eastern or northeastern plants collected in this region may be
mentioned:

Viola palustris.　　　　　　　　　　　　*Hypericum canadense.*
Viola blanda.　　　　　　　　　　　　　*Tetragonanthus deflexus.*
Lobelia spicata hirtella.　　　　　　　　*Fragaria virginiana.*
Stachys aspera.　　　　　　　　　　　　*Solidago erecta?*

The most remarkable "find," however, was that of the true *Aquilegia
brevistyla* in the United States. The Rocky Mountain plant, so named,
proves to be a distinct species and has received the name *A. saximontana.*

As I have said before, the valleys are rich grass land. Even the
dryer ones furnish a good pasture and along the water courses are
excellent hay lands. One of the men accompanying the geological survey under Jenney, named "California Joe," expressed himself, "There's
gold from the grass roots down but there's more gold from the grass
roots up." Around Custer, the place to which the first great rush of
gold hunters was directed, stock raising or farming seems to be more
profitable than gold digging.

In a meadow near French Creek the grass stood 1 meter high. The
most common grasses were:

Panicularia nervata.　　　　　　　　　*Calamagrostis canadensis.*
Agrostis alba.　　　　　　　　　　　　*Calamagrostis dubia.*
Poa nemoralis.　　　　　　　　　　　　*Agropyron repens glaucum.*
Alopecurus geniculatus fulvus.

In a slough I found *Spartina cynosuroides, Beckmannia crucæformis,*
and *Panicularia americana.* In a glen below Sylvan Lake were found

two eastern grasses, *Oryzopsis juncea* and *O. asperifolia*. Near the railroad occurred two forms of *Poa nevadensis*, and *Bromus pumpillianus*, both of a more western range. On a wooded hill, together with the three common Stipas. *S. spartea*. *S. comata*, and *S. viridula*, grew a fourth. *S. richardsonii*, of a more western range, and also *Danthonia spicata*, from the East.

But the most peculiar feature of this region is the damp atmosphere. The Harney Range differs in that respect from the Northern Hills. On account of this dampness, and differently from mountain regions in general, the Harney Range abounds in lichens, liverworts, mosses, and ferns, especially on the north side of the crags, where the rocks in many places are literally covered by lichens and the base and crevices lined by mosses and ferns. The lichens and mosses were collected only incidentally, but a good collection of ferns was made. My list contains the following from this region:

Polypodium vulgare.	*Pteris aquilina.*
Asplenium trichomanes.	*Asplenium septentrionale.*
Asplenium filix-femina.	*Dryopteris filix-mas.*
Phegopteris dryopteris.	*Cystopteris fragilis.*
Woodsia oregana.	*Woodsia scopulina.*
Botrychium matricariæfolium?	*Selaginella rupestris.*
Polypodium vulgare rotundatum.	

LIMESTONE DISTRICT.

The Limestone District is a high table-land, running from south to north, on the Wyoming line. It is separated from the Harney Range and the other hills by a valley. This table-land is the watershed of the Black Hills, giving rise to Spearfish, Rapid, French, and Red Canyon creeks on the east side, and Red Water, Inyankara, and Beaver creeks on the west side. The plateau is 1,800 to 2,000 meters high, the highest point, Crooks Tower, being, next to Harney Peak, the highest in the hills. The surface is made up of pine-covered ridges running north and south. The valleys between these ridges are composed of excellent hay land. The region resembles much some parts of Sweden. The pine-covered hills were here, so also the meadows with the knee-deep grass, and the flowers were in greater profusion and greater variety of color than I have seen elsewhere in America. The Swedish species were seldom present, but they had their counterparts: *Hieracium*, *Scorzonera*, and *Hypochœris* were matched by *Rudbeckia hirta*, *Gaillardia aristata*, and *Helianthus maximiliani*; *Lathyrus* and *Vicia* by *Lupinus parviflorus* and *L. sericeus*; *Geranium sylvaticum* by *Geranium richardsonii*; *Chrysanthemum leucanthemum* by *Aster ptarmicoides*; and *Solidago virgaurea* by *Solidago missouriensis*. In the border of the woods the same old *Epilobium angustifolium* presented itself.

The only trees seen in the district were the pine and the quaking aspen.

Of shrubs were observed:

Salix bebbiana.
Ribes cereum.
Shepherdia canadensis.
Ceanothus fendleri.

Salix discolor.
Juniperus communis.
Elaeagnus argentea.

The Ceanothus has hitherto been reported only from southern Colorado and southward.

Other remarkable plants were:

Epilobium hornemanii.
Helianthella quinquenervis.
Astragalus convallarius.
Pellaa breweri.

Balsamorhiza sagittata.
Frasera speciosa.
Epilobium drummondii.
Lupinus sericeus.

all from a more western or southern range.

To me this region looked as promising as any in the Black Hills for agricultural purposes. As said before, the valleys were excellent hay lands. The grasses were about the same as those around Custer. The dryer valleys and the woods would furnish enough of summer pasture. During the winter the stock must be fed with hay as the snowfall is very heavy. Sometime after I had visited the region I heard that this was the principal reason why many of the squatters had left the region. The soil was a black loam containing a considerable amount of lime, the valleys were less rough than those of the parks of the preceding region and could easily be made into fields.

NORTHERN HILLS.

The Northern Hills, notwithstanding their great height, look more like hills than mountains. Even the highest, as Terrys Peak, Custers Peak, etc., are covered with woods to the top. The larger part of the region is the northern half of the Archaean formation. As said before this is composed of comparatively soft slates and schists. The rivers have worn out deep canyons, many volcanic eruptions have thrown up cones of igneous rocks, and the remainders here and there of the broken strata of Potsdam sandstone and Carboniferous limestone make the country more uneven. The woody flora resembles that of the Harney Range, but the pine is more predominant. The elm is lacking in this region and the oak is confined to the foothills and neighboring canyons. The following shrubs and climbers may be added.

Ceanothus ovatus.
Potentilla fruticosa.
Vitis vulpina.

Viburnum lentago.
Lonicera hirsuta glaucescens.
Parthenocissus quinquefolia.

The whole region seems to have been one large pine forest; but now large tracts are made bare by the ravages of lumbermen, mining companies, fire, and cyclones, nothing being left but stumps, fallen logs, and the underbrush. The second forest will consist of deciduous trees, as aspen, willows, birch, and cherry. The mining resources of the Hills, especially around Lead City and Deadwood, are well known. The

Black Hills and Fort Pierre Railroad was built by the Homestake Mining Company, principally for the purpose of transporting wood and lumber to their mines and stamp mills, and other roads have been built by other companies. Sawmills are scattered all over the Hills, and it will be no wonder if in a short time the dark pine forest is gone and the name " Black Hills" has become meaningless.

The valleys of this region are very narrow, and in that small part in which I collected, little of their natural condition was left. The Elk Canyon in many places was just wide enough to give room for the creek and the railroad. The nearly perpendicular sides were as much as 200 to 300 meters high. Around Lead City and Deadwood railroads and wagon roads wind through the narrow valleys, and the small patches of grass left are well cropped down by the town cows. At Rochford only I found a good meadow. The grasses were the same as in the other regions of the hills, but the blue grasses were more common. The following grasses may be mentioned as of special interest:

Oryzopsis micrantha.	*Arena striata.*
Bromus pumpellianus.	*Elymus dasystachys.*

from Elk Canyon.

Calamagrostis sylvatica americana.	*Agropyron violaceum majus.*
Panicum depauperatum.	*Festuca ovina.*

from the neighborhood of Lead City.

The Northern Hills, especially the canyons, contain more Eastern as well as Western plants than any other part of the hills. Among those not given in Coulter's Manual, which is supposed to cover all the territory west of the one hundredth meridian, are:

Viola scabriuscula.	*Polygala senega latifolia.*
Lathyrus ochroleucus.	*Naumburgia thyrsiflora.*
Tetragonanthus deflexus.	*Lappula deflexa americana.*
Lappula virginiana.	

Of Western plants were found:

Thalictrum occidentale.	*Thalictrum venulosum.*
Claytonia perfoliata amplectens.	*Lupinus parviflorus.*
Lupinus sericeus (?).	*Spiraea caespitosa.*
Potentilla glandulosa.	*Heuchera parvifolia.*
Epilobium drummondii.	*Osmorrhiza nuda.*
Arnica cordifolia.	*Arnica alpina.*
Hieracium fendleri.	*Vaccinium myrtillus microphyllum.*
Frasera speciosa.	*Mertensia sibirica.*
Mimulus luteus.	*Calochortus gunnisoni.*
Potentilla humifusa.	*Lesquerella spatulata.*

Among the most interesting finds was a patch of caraway, *Carum carui*, which I found in the wilderness 3 or 4 miles north of Deadwood. Perhaps some German or Scandinavian gold hunter had happened to drop a piece of old country cheese spiced with the customary caraway seed, and hence the patch.

GENERAL REMARKS.

From the foregoing can be seen what a varied flora the Black Hills have. There are found plants from the East, from the Saskatchewan region, from the prairies and table-lands west of the Missouri River, from the Rocky Mountains, and even from the region west thereof. In the foothills and the lower parts of the Hills proper the flora is essentially the same as that of the surrounding plains, with an addition of Eastern plants which have ascended the streams. In the higher parts the flora is more of a Northern origin. Most of the plants composing it are of a more or less transcontinental distribution but often characteristic of a higher latitude. Some can be said to belong to the Rocky Mountain region. The only trees of Western origin are *Pinus ponderosa scopulorum*, and *Betula occidentalis;* the others are Eastern or transcontinental. The flora resembles therefore more that of the region around the Great Lakes than that of the Rockies.

The collection contains a little over 700 Phænogams and Fernworts. This is certainly far from all that grow in the region. A few more known to occur in the Black Hills could have been added to the list, as for instance, *Mentzelia oligosperma* and *Ilysanthes gratioloides*, collected by Mr. A. F. Woods; *Onoclea sensibilis* and *Aster salsuginosus*, by Prof. T. A. Williams; *Fritillaria linearis*, by Miss Pratt, of Piedmont; and *Sorbus sambucifolia*, by Mr. Runkel, the owner of the sawmills at Runkels. *Viburnum prunifolium* was also reported by a physician of Custer, but perhaps *V. lentago* was mistaken for it. A squatter told me that he had cut hickory poles on the Squaw Creek, a statement which seems doubtful. Jenney, in his report on the Geological Survey of the Black Hills, reports the black spruce and mulberry as growing in the hills. The former probably was confounded with the white spruce, and the occurrence of the latter needs verification.

To the following botanists acknowledgements are due for help in the determination of the species. The Carices have been determined by Prof. L. H. Bailey, the genera Epilobium and Gayophytum by Dr. William Trelease, Polygonum by Mr. J. K. Small, Salix by Mr. M. S. Bebb. The determinations of Juncaceæ, Gramineæ, and Umbelliferæ have been verified by Mr. Frederick V. Coville, Prof. F. Lamson-Scribner, and Mr. J. N. Rose, who have also made a few corrections where needed. The description of *Poa pseudopratensis* is drawn by Professor Scribner.

In the identification of the collection, the plants have been compared with specimens in the National Herbarium and the herbarium of the University of Nebraska. Thanks are also due to Prof. N. L. Britton, of Columbia College, and Prof. John Macoun, of Ottawa, Canada, for the loan of specimens for comparison.[1]

[1] The acknowledgments expressed, p. 148 of this volume, footnote 1, are also here renewed.

CATALOGUE OF SPECIES.

RANUNCULACEÆ.

Clematis scottii Porter; Port. & Coult. Fl. Col. 1 (1874).

The specimens in this collection are like those collected by Dr. Scoville in Colorado, but not like those obtained by Lemmon in Arizona, which undoubtedly belong to a distinct species. Coulter, in the Manual of the Rocky Mountain Region, describes the sepals as less hairy than those of *C. douglasii*. In mine they are fully as hairy, but thicker and shorter.

On hillsides near Hot Springs, altitude 1,060 m., June 1, 17; in fruit, August 2 (No. 481).

Clematis ligusticifolia Nutt.; Torr. & Gray, Fl. i, 9 (1838).

In canyons among the foothills: Hot Springs, altitude 1,050 m., August 2 (No. 182). A form with large (5 to 7 cm. long) and dullish leaflets and very long (15 to 22 cm.) and slender peduncles, was collected on the very steep sides of Hot Springs Canyon, near the Chautauqua grounds, altitude 1,050 m., August 3 (No. 483).

Clematis alpina tenuiloba (Gray).

Gray[1] makes this a subvariety of *C. alpina occidentalis* Gray, which is described as having smooth achenes. In my specimens they are silky. The plant further differs from *C. alpina ochotensis* or *occidentalis* in having more delicate stems, smaller and more lobed leaves with more rounded lobes and sinuses, and longer, lanceolate sepals. This plant has been collected also by Dr. Chas. E. Bessey at Maniton, Colo. It was sold in albums of Black Hills flowers at Deadwood under the name of *C. douglasii*. Perhaps the silkiness of the achenes was the cause of this error.

Here and there in canyons, in the Black Hills proper, near Piedmont, altitude 1,200 m., June 27; Lead City, altitude 1,600 m., July 6; Bull Springs, in the Limestone District west of Custer, altitude 1,900 m., July 27 (No. 484).

Pulsatilla hirsutissima (Pursh) Britton, Ann. N. Y. Acad. vi, 217 (1891); *Clematis hirsutissima* Pursh, Fl. i, 385 (1814).

Common in the Hills: Custer, altitude 1,650 m., May 28, 31, June 4 (No. 485).

Anemone multifida Poir. Encyl. Suppl. i, 364 (1810).

Not uncommon in the Northern Hills: Elk Canyon, altitude 1,200 m., June 29; Lead City, altitude 1,800 m., July 4; Rochford, altitude 1,700 m., July 11 (No. 486).

Anemone cylindrica Gray, Ann. Lyc. N. Y. iii, 221 (1836).

In the Northern Hills: Hermosa, altitude 1,000 m., June 21; Lead City, altitude 1,700 m., July 6; Rochford, altitude 1,650 m., July 11 (No. 487). Several of the plants have some of the peduncles with secondary involucres. In a few specimens from Elk Canyon, altitude 1,200 m., June 29, the divisions are also broad, and the plants can not be distinguished from *A. virginiana*, except by the very short style (No. 488).

Thalictrum purpurascens L. Sp. Pl. i, 546 (1753).

In canyons, among the foothills: Hot Springs, altitude 1,050 m., June 18; Elk Canyon, altitude 1,200 m., June 29 (No. 189).

Thalictrum occidentale Gray, Proc. Amer. Acad. viii, 372 (1872).

The specimens are rather too young for identification. None were seen in fruit, as I was not in the locality of the plant except in the early part of the summer. The foliage is very like that of *T. occidentale*, and the plant agrees well with the description of that species. If the determination is correct, the range of *T. occidentale* is extended far east.

Elk Canyon, altitude 1,200 m., June 29 (No. 490).

Thalictrum venulosum Trelease, Proc. Bost. Soc. Nat. Hist. xxiii, 302 (1886).

This is *T. dioicum* L., of Newton & Jenney's Report.[2]

The only specimens in fruit seem to be typical. They were collected near Bull

---

[1] In Newton & Jenney, Geol. Surv. Black Hills, 531 (1880).   [2] Loc. cit., p. 532.

Springs, altitude 1,000 m., July 27. Younger specimens from Little Elk Canyon, altitude 1,100 m., June 29, have somewhat larger and thinner leaves (No. 491).

Batrachium divaricatum (Schrank) Wimm. Fl. Schles. 10 (1841); *Ranunculus divaricatus* Schrank, Baier. Fl. ii, 104 (1789).

The specimens resemble those of my Nebraska collections, except that the peduncles are much shorter.

In brooks: Beaver Creek, near Buffalo Gap, altitude 975 m., June 21; Rapid Creek, above Rapid City, altitude 1,000 m., June 25; Elk Creek, altitude 1,100 m., June 28 (No. 492).

Cyrtorhyncha cymbalaria (Pursh) Britton, Mem. Torr. Club, v. 161 (1894); *Ranunculus cymbalaria* Pursh, Fl. i. 392 (1814).[1]

Around springs; common: Hot Springs, altitude 1,050 m., June 10 (No. 493).

Ranunculus cardiophyllus Hook. Fl. Bor. Amer. i. 11 (1829).

This is not included in Coulter's Manual. The range is hence extended westward. The petals in my specimen are broadly ovate, large, bright yellow, the sepals very pubescent.

Custer, altitude 1,625 m., June 4; Rochford, altitude 1,600 m., July 11 (No. 494).

Ranunculus ovalis Raf. Proc. Dec. 36 (1814).

Not uncommon in shady places: Custer, altitude 1,625 m., May 28. Some specimens resemble somewhat the preceding species in size and habit, but the petals are oblong-rhombic (No. 495).

Ranunculus abortivus L. Sp. Pl. i, 551 (1753).

All specimens collected in the Hills are very slender and with thin leaves. This is especially the case with those from Elk Canyon, altitude 1,200 m., June 29 (No. 496). Those from Runkels, altitude 1,300 m., June 29 (No. 497), are more stout and approach the ordinary form.

Ranunculus sceleratus L. Sp. Pl. i, 551 (1753).

In and near streams: Piedmont, altitude 1,000 m., June 27; Elk Creek, altitude 1,200 m., June 29; Rochford, altitude 1,600 m., July 11 (No. 498).

Ranunculus pennsylvanicus L. f. Suppl. 272 (1781).

Wet places; common: Lead City, altitude 1,500 m., July 6; Custer, altitude 1,650 m., July 15 (No. 499).

Ranunculus macounii Britton, Trans. N. Y. Acad. xii, 3 (1892).

Hooker, in the Flora Boreali-Americana, describes his *R. hispidus* as being erect. Dr. Britton, loc. cit., says:

"This is a spreading or trailing species, not stoloniferous as far as I know." As far as I can judge, there are two forms of this species; one, generally ascending, but sometimes erect, sometimes even spreading, the other widely trailing and stoloniferous. All the specimens from the Black Hills were of the former character, and may be regarded as the typical *R. macounii*, as they answer best the description of *R. hispidus* Hooker, on which *R. macounii* was based.

Very common throughout the Black Hills: Hot Springs, altitude 1,050 m., June 15; Hermosa, altitude 1,050 m., June 23; Lead City, altitude 1,500 m., July 6 (No. 500).

Aquilegia canadensis L. Sp. Pl. i, 533 (1753).

Common: Rapid City, altitude 1,000 m., June 25; Little Elk Canyon, altitude 1,100 m., June 28; Rochford, altitude 1,600 m., July 11 (No. 501).

Aquilegia canadensis formosa (Fisch.) Cooper, Pac. R. Rep. xii, 55 (1860); *Aquilegia formosa* Fisch.: DC. Prodr. i, 50 (1824).

This seems to grade into the preceding, from which it differs in the shorter spur and longer sepals, which are about twice the length of the petals. In my specimens the leaves are larger and more glaucous.

Rare: Elk Canyon, altitude 1,200 m., June 29 (No. 502).

---

[1] See remarks on the synonymy of this species, this volume, p. 148 (1895).

PLATE XVIII.

AQUILEGIA BREVISTYLA Hook.

Aquilegia brevistyla Hook. Fl. Bor. Amer. i, 24 (1829).   Pl. XVIII.

This plant is very rare in the United States. Unless the locality given in the sixth edition of Gray's Manual[1] belongs to this plant, the station given below, as far as I know, is the first one recorded in the country. All specimens I have seen from the Rocky Mountains belong to another species, which I have named *A. saximontana*.[2] The original description of *A. brevistyla* is as follows: "Subpubescens, calcaribus incurvis limbo brevioribus, stylis brevioribus inclusis, staminibus corolla subrevioribus." To this Hooker adds, in smaller type: "Caulis foliaque fere omnino ut in *A. vulgare*. Flores duplo minores, caerulei plerumque pubescentes." *"Pistilla* 5. Germina lineari-cylindracea, pubescentia, in stylis apice leniter recurvis sensim attenuata, staminibus brevioribus. Capsulae 5, unciam longae, in stylo brevi vix duas lineas longo terminatae."

This description does not fit the Rocky Mountain plant, as in it neither the stem nor the flower nor the capsule is pubescent, but the plant is perfectly smooth. Neither does the stem nor the leaves resemble those of *A. vulgaris*. The Rocky Mountain plant is more or less cespitose, with many low (1 to 2 dm. high) stems from the caudex, which is covered with old leafstalks. In *A. vulgaris* the stem is tall (4 to 10 dm. or more high) and simple. The leaves are of a firm texture in the latter, the root leaves long-petioled and twice-ternate, the stem leaves on short petioles or subsessile, often simply ternate or simple and 3-lobed. In the Rocky Mountain plant the leaves are thin, all slender-petioled and twice ternate, the upper, however, sometimes reduced. The plants of my collection and specimens of *A. brevistyla* from western British America very much resemble *A. vulgaris*, but differ in their shorter styles, the smaller size of the flowers, and the form of the corolla. In *A. vulgaris* the limb is truncate or retuse, much shorter than the spur, and generally shorter than the stamens. In *A. brevistyla* the limb is oblong, truncate, longer than the short spur and the stamens. The corolla, peduncles, upper part of the stem, and the capsules are in the specimens mentioned, as they should be according to the original description, viz. pubescent.

In nearly all the literature in this country in which *A. brevistyla* is mentioned, the reference is to *A. saximontana* instead. Torrey & Gray's Flora is an exception. Here the description is essentially the same as in Hooker's Flora. In both the distribution of the species is given as "Western part of Canada, as far north as Bear Lake," Gray's Manual, sixth edition, perhaps includes both. All the other descriptions I have seen refer to the Rocky Mountain plant. The best one is given by Marcus E. Jones.[3] This I shall use as the basis for my description of *A. saximontana*, adding such characters as will better show the distinction between this and *A. brevistyla*. Even a comparison between Jones's description (or Porter's in Flora of Colorado, or Coulter's in Manual of Rocky Mountain Region), and the original one in Hooker's Flora will show that they are drawn from different plants.[4]

[1] Gray, Man. ed. 6, 46 (1890).

[2] See page 482, in footnote.

[3] Zoe, iv, 258, October, 1893.

[4] The North American species of Aquilegia with curved spur may be disposed in the following way:

A. *Stem 4 to 10 dm. high.*

a. *Style in fruit more than 1 cm. long.*

A. VULGARIS L. Sp. Pl. i, 533 (1753).
Limb of the corolla shorter than the spur and the stamens; flowers blue, red, or white. Escaped from gardens.

A. FLAVESCENS Wats. Bot. King. Surv. v, 10 (1811).
Limb of the corolla of the length of the spur but shorter than the stamens;

On dark, wooded hillsides; rare: Little Elk Canyon, altitude 1,200 m., June 28; Oreville, altitude 1,650 m., July 16 (No. 503).

Delphinium bicolor Nutt.; Torr. & Gr. Fl. i, 33 (1838).

Variable. In the collection there are three forms, which probably belong here. One is 4 to 6 dm. high, with most leaves near the base, more or less glandular-pubescent throughout, even to the pods. Custer, altitude 1,650 m., June 3, Aug. 1 (No. 504).

Another form is like this, but perfectly smooth and with thinner sepals; in some specimens the flowers are purplish pink. Runkels, altitude 1,300 m., June 30 (No. 505).

The third is a tall form 7 to 10 dm. high, glandular-pubescent, and with broader, more pointed divisions to the leaves. It is the same as *D. menziesii utahense* Wats.[1] Elk Canyon, altitude 1,200 m., June 29 (No. 1205).

Aconitum fischeri Reich. Monogr. Gen. Acon. i, 22 (1820).

The common American form is a tall plant, generally 1 to 1.5 m. high, robust, pubescent, and viscid. The divisions of the leaves in my specimens, as well as in some corolla yellowish, sometimes tinged with red or blue. See M. E. Jones. loc. cit. In the Wasatch Mountains from Utah to British America.

b. *Style in fruit 5 to 7 mm. long.*

A. **brevistyla** Hook. Fl. Bor. Amer. i, 24 (1829); Torr. & Gr. Fl. i, 30; Walp. Rep. i, 51, etc. (Some of the other references in Wats. Bibl. Index, p. 6, may belong here, as *A. vulgaris?* Richards. App. Frankl. Journ. 740).

Stem 1 to 10 dm. high, simple, pubescent, or glandular above, especially on the peduncles and flowers: root leaves 2-ternate on *stout* petioles; stem leaves diminishing upward, often ternate and short-petioled or the upper simple, 3-lobed and sessile; pedicels stout and recurved; sepals blue, acute; limb of the *petals yellowish white, longer than the blue, curved spur and the stamens; ovary* pubescent: pod 2 to 2.5 cm. long, reticulate and glandular-pubescent.

Western Canada, Red River Valley (?), and the Black Hills. Specimens seen: Canada, Morley (Albertina), 1885, John Macoun; McKenzie River, Louis Anderson; South Dakota, No. 503 of this collection.

B. *Stem 0.5 to 2 dm. high, subcespitose.*

a. *Style in fruit about 0.5 cm. long.*

A. **saximontana** Rydberg; Gray, Syn. Fl. i, pt. 1, 43 (1895); *A. vulgaris brevistyla* Gray, Amer. Journ. Sci. ser. 2. xxxiii, 410, and Proc. Acad. Phila. 1863, 57 (1863), name only; Porter, Port. & Coult. Fl. Col. 4 (1874), description; *A. brevistyla* Coulter, Man. Rock. Mount. Reg. 10 (1885); Jones. Zoe, iv, 258 (1893). PL. XIX.

Stem 1 to 2 dm. high, densely tufted, scarcely exceeding the leaves, perfectly smooth; leaves twice-ternate, all on slender petioles thin, the upper a little smaller; leaflets 8 to 15 mm. long, with long petiolules, pedicels slender, upright; sepals greenish and obtuse or blue and acute; limb of the petals yellow, longer than the blue, curved spur, and the stamens and pistils; ovary smooth: pod 1.5 to 2 cm., smooth.

Rocky Mountains of Colorado. Specimens examined: Colorado, Dr. James (labeled A. caerulea, var. ?); 1861, C. C. Parry, No. 90; 1862, Hall & Harbour, No. 23; 1860, Scoville; Argentine Pass, 1878, M. E. Jones, No. 875; Gray's Peak, 1895, P. A. Rydberg and C. L. Shear.

b. *Style in fruit about 1 cm. long.*

A. **jonesii** Parry, Amer. Nat. no. 8, 211.

Cespitose, 0.5 to 1 dm. high; leaflets about 0.5 cm. long, nearly sessile; spur nearly straight.

Rocky Mountains of Wyoming and Montana.

[1] Bot. King. Surv. 112 (1871).

AQUILEGIA SAXIMONTANA Rydberg.

from Colorado and Wyoming, are much narrower than in those from the Pacific Slope. Near Terrys Peak, altitude 1,900 m., July 6; Rochford, altitude 1,700 m., July 12 (No. 506).

Specimens, collected among peat moss, below Sylvan Lake, 4 miles south-southwest from Harneys Peak, altitude 2,000 m., July 18 (No. 507) are much smaller, 3 to 6 dm. high, slender, few-flowered, less pubescent, with finer lobes to the leaves and bluer flowers, the hoods of which are more semicircular in outline.

Actæa spicata rubra Ait. Hort. Kew. ii, 221 (1789).

Professor Greene regards this as specifically distinct from *A. spicata.* Perhaps it is so, but the characters pointed out which are to separate *A. spicata* from *A. rubra* are not constant, at least in Scandinavian specimens of the former. Even the fruit is sometimes bright red in them.

Rare: Elk Canyon, altitude 1,200 m., June 29 (No. 508).

Actæa spicata arguta (Nutt.) Torr. Pac. R. Rep. iv, 63 (1856); *Actæa arguta* Nutt.; Torr. & Gr. Fl. i, 35 (1838).

This often has the fruit white and much larger and more elongated than in the red-fruited variety. Perhaps they are distinct, but I could not find any other character that would separate them.

Little Elk Canyon, altitude 1,100 m., July 18; Custer, altitude 1,650 m., August 15 (No. 509).

BERBERIDACEÆ.

Berberis aquifolium Pursh, Fl. i, 219 (1814); *Berberis repens* Lindl. Bot. Reg. t. 1176 (1828).

Without doubt this is the original *Berberis aquifolium* Pursh, and Lindley made a mistake when he supposed that the name belonged to the taller species of the Columbia River basin. Lindley's statement that Pursh's drawing was made from Menzies's plant, that is, the *B. aquifolium* [1] of Hooker and of Lindley, is evidently wrong, as Pursh does not cite Menzies as having collected it. The plate was made from a specimen of Lewis's collection, and it as well as the description shows that the plant belongs to what has been known as *B. repens* Lindley.[2] Sweet, in British Flower Garden, says: "Mr. Lindley's observations on *B. aquifolium* are wrong; the very specimen figured by Pursh is now in his herbarium in Mr. Lambert's collection; the name *B. repens* published in the Bot. Reg. must therefore be disused."

In canyons: Hot Springs, altitude 1,100 m., June 12; Little Elk, altitude 1,200 m., June 28 (No. 510).

PAPAVERACEÆ.

Argemone alba Lestib. Bot. Belg. ed. 2, iii, 133 (1799); *A. albiflora* Hornem. Hort. Hafn. 489 (1813-15).[3]

In draws among the foothills. Hermosa, altitude 1,025 m., July 24; 15 miles east of Custer, altitude about 1,400 m., July 23 (No. 511).

FUMARIACEÆ.

Capnoides aureum (Willd.) Kuntze, Rev. Gen. Pl. i, 14 (1891); *Corydalis aurea* Willd. Enum. 740 (1809).

Rare: Elk Canyon, on the railroad embankment, altitude 1,200 m., June 29 (No. 512).

---

[1] This must take the name *Berberis nutkana* (DC.) Kearney, Trans. N. Y. Acad. xiv, 29 (1894); *Mahonia aquifolium nutkana* DC. Syst. ii, 20 (1821).

[2] One leaflet in Pursh's figure (fig. 1) may belong to *B. aquifolium* Hook.; at least this was Watson's view.

[3] See my notes, p. 149 of this volume. Mr. Prain has shown (Journ. Bot. xxxiii, 329) that *P. albiflora* is antedated by *P. alba.* Both are based on specimens from the Southern States. Mr. Prain thinks that the plant of the Western plains is distinct, and names it *A. intermedia.* I can not, however, find any constant character that will separate the two.

Capnoides curvisiliquum (Engelm.) Kuntze, Rev. Gen. Pl. ii, 14 (1891); *Corydalis curvisiliqua* Engelm.; Gray, Man. ed. 5, 62 (1867).
This is not uncommon in the Black Hills: Sylvan Lake, altitude 1,900 m., June 8; Hot Springs, altitude 1.050 m., June 13 (No. 513).

NYMPHÆACEÆ.

Nymphæa advena Soland. in Ait. Hort. Kew. ii, 226 (1789). Leaves of this species were seen in Rapid Creek, 6 miles above Rapid City, but no specimens were secured.

CRUCIFERÆ.

Draba carolinana micrantha (Nutt.) Gray, Man. ed. 5, 72 (1867); *Draba micrantha* Nutt.; Torr. & Gr. Fl. i, 109 (1838).
Barren hills, rare: Hot Springs, altitude 1.100 m., June 13 (No. 514).

Draba nemorosa L. Sp. Pl. ii, 643 (1753).
The specimens of this collection are pubescent even to the pods, and may belong to the variety *hebecarpa* Lindl., but the hairy form has been regarded, by De Candolle and others, as the typical *D. nemorosa* L.
Early in the spring: Custer, altitude 1,650 m., June 1 (No. 515).

Draba aurea Vahl. in Hornem, Fors. Dansk. Œcon. Pl. ed. 2, 599 (1806).
My specimens differ from the common form in being more slender, and in having entire, thin leaves, smaller, paler petals with more slender claw, and longer, decidedly twisted pods. The peduncles and pedicels are ciliate and the sepals narrow. I took it for the variety *stylosa*. It resembles somewhat the original specimens of this, from Fendler's collection: but the pods are much longer and the style shorter. I do not wish to give it a varietal name, as I have specimens from only one locality.
In a shady place, at the foot of a high cliff, below Sylvan Lake, altitude 2,000 m., July 18 (No. 516).

Arabis glabra (L.) Bernh. Verz. Syst. Erf. 195 (1800); *Turritis glabra* L. Sp. Pl. ii, 666 (1753).
Rare: Along the railroad in Elk Canyon, altitude 1,200 m., June 29 (No. 517).

Arabis hirsuta (L.) Scop. Fl. Carn. ed. 2, ii, 30 (1772); *Turritis hirsuta* L. Sp. Pl. ii, 666 (1753).
Common: Custer, altitude 1,650 m., May 30, June 3; Hot Springs, altitude 1,100 m., June 13; Elk Canyon, altitude 1,200 m., June 29. The specimens from the latter place are unusually slender (No. 518).

Arabis holbœllii Hornem. Fl. Dan. xi, t. 1879 (1827).
The flowers in my specimens are seldom reflexed; the pods are a little curved and reflexed.
Common: Custer, altitude 1,650 m., June 5; Elk Canyon, altitude 1,200 m., June 29; Lead City, altitude 1,500 m., July 6 (No. 519).

Arabis holbœllii retrofracta (Graham); *Arabis retrofracta* Graham, Edinb. Phil. Journ. 344 (1829).
The latter has generally been regarded as a synonym of *A. holbœllii*. There seem, however, to be two or three different forms of this species, one of which has more slender pods, which are perfectly straight, and smaller flowers. This agrees with the description of *Turritis retrofracta* in Hooker's Flora Boreali-Americana, Volume I, page 44. The root leaves are spatulate, petioled, the stem leaves sessile, lanceolate, with a hastate, half-clasping base.
Elk Canyon, altitude 1,200 m., June 29 (No. 520).

Stanleya pinnata (Pursh) Britton. Trans. N. Y. Acad. viii, 62 (1889); *Cleome pinnata* Pursh, Fl. ii, 739 (1811).
On prairie, 1 mile east of Fall River Falls, altitude 1,000 m., June 18 (No. 521).

Erysimum asperum (Nutt.) DC. Syst. ii, 505 (1821); *Cheiranthus asper* Nutt. Gen. ii, 69 (1818).

Hot Springs, altitude 1,050 m., June 14; Hermosa, altitude 1,000 m., June 23; Rapid City, altitude 1,000 m., June 26 (No. 522).

A form with slender, twisted pods was collected on the hills north of Deadwood, altitude 1,500 m., July 5 (No. 523).

Erysimum cheiranthoides L. Sp. Pl. ii, 661 (1753).

Hot Springs, altitude 1,050 m., June 14; Rochford, altitude 1,050 m., July 11 (No. 524).

Erysimum inconspicuum (Wats.) MacMillan, Metasp. Minn. Val. 268 (1892); *Erysimum asperum inconspicuum* Wats. Bot. King Surv. 24 (1871).

This resembles very much *E. virgatum* Roth of Europe, and perhaps is only its American form. It was growing together with *E. asperum* and *E. cheiranthoides*, and in the field it seemed as if possibly it might be a hybrid of the two. In pubescence, color, and form of the flowers, and form of the pod it is more like *E. asperum;* the general habit is that of *E. cheiranthoides,* and the size of the flowers intermediate between those of the two.

Hot Springs, altitude 1,050 m., June 13 (No. 525).

Sisymbrium pinnatum (Walt.) Greene, Bull. Cal. Acad. ii, 390 (1887); *Erysimum pinnatum* Walt. Fl. Car. 174 (1788).

This is very variable. Some specimens are more or less canescent and have the seeds plainly in two rows (No. 526); others are smooth or, on the upper parts, glandular and have the seeds apparently in one row, characters that should belong to *S. incisum* Engelm. (No. 527). The two were growing together at Hot Srpings, altitude 1,075 m., June 14.

Brassica alba (L.) Boiss. Voy. Espagne, ii, 239 (1839–45); *Sinapis alba* L. Sp. Pl. ii, 668 (1753).

Railroad embankment, Buffalo Gap, altitude 991 m., June 21 (No. 529).

Brassica sinapistrum Boiss. Voy. Espagne, ii, 39 (1839–45).

Buffalo Gap, altitude 991 m., June 21 (No. 528).

Roripa palustris (L.) Bess. Enum. 27 (1821); *Sisymbrium amphibium palustre* L. Sp. Pl. ii, 657 (1753).

Rare in the region: Rapid City, altitude 1,000 m., June 26 (No. 530).

Roripa nasturtium (L.) Rusby, Mem. Torr. Club. iii, 5 (1893); *Sisymbrium nasturtium* L. Sp. Pl. ii, 657 (1753).

Fall River, near the Chautauqua grounds, above Hot Springs, altitude 1,050 m., June 14 (No. 531).

A form, very low, creeping, extensively rooting, with very fleshy leaves of 0 to 2 small pairs of leaflets and a larger, nearly orbicular, odd leaflet, and with short and thick pods, was growing in the warm springs, at Hot Springs, altitude 1,050 m., June 14 (No. 532).

Lesquerella argentea arenosa (Richards.) Wats. Proc. Amer. Acad. xxiii, 252 (1888); *Vesicaria arenosa* Richards. Bot. App. 713 (1823).

There are no specimens of this variety in the National Herbarium, but I think it is this plant (which is illustrated by specimens in the Harvard Herbarium) that Dr. Watson had in view in his revision. The form of the leaves does not agree fully with the original description in the Appendix to Franklin's Journal, being narrower and with entire margin. The figure of *V. arctica* in Curtis's Botanical Magazine,[1] which, according to Watson, is the same as *V. arenosa* Richards., is drawn from a young specimen, but resembles this much, although it seems to be a little stouter

My plant is densely stellate-pubescent, much branched from the perennial caudex; radical leaves broadly spatulate or oblanceolate, often a little acutish; stem 1.5 dm.

---

high; pod globose or a little elongated, stellate-pubescent with a long style, pedicels recurved as in *L. ludoriciana*, which it resembles, although it is more slender and more branched and has smaller pods.

Hillside, Hot Springs, altitude 1,100 m., June 11 and August 3 (No. 533).

Lesquerella spatulata sp. nov.

Low and somewhat caespitose; stems 3 to 10 cm. high, subscapose; leaves all radical, spatulate or oblanceolate, decurrent on the petiole; petals spatulate, yellow; pods on recurved pedicels, ovate, only slightly compressed toward the apex, finely pubescent, 1-seeded; septum not perforate; style scarcely as long as the mature pod.

Like the last in habit, but lower. The pod, however, is ovate, pointed, and slightly compressed toward the apex, about the length of the slender style but shorter than the pedicel, which is reflexed and then curved upward. The plant is somewhat intermediate between *L. montana*, the preceding species, and *L. alpina*. From *L. montana* it differs in its more slender habit, narrower leaves, and smaller pods; from *L. alpina*, in being much taller and in having broader leaves, less compressed pods, longer and recurved pedicels, and shorter style. In *L. alpina* the style is longer than the pod, the pedicels erect, and the septum perforated. It differs from *L. argentea arenosa* in the form of the pod. Similar specimens were found in the Harvard Herbarium, viz, in C. C. Parry's collection of 1873 (No. 21) and in the collection of Jenney's expedition, 1875. These were referred to *L. alpina* by Dr. Watson. In the Columbia College herbarium there are the following specimens: H J. Webber, from Belmont, Nebraska, 1889; Dawson, Milk River, N. W. T., 1883. Dry hilltop, north of Deadwood, altitude 1,600 m., July 5 (No. 531).

Camelina sativa (L.) Crantz, Stirp. Austr. i, 18 (1762); *Myagrum sativum* L. Sp. Pl. ii, 641 (1753).

Introduced: Railroad embankment above Custer, altitude 1,700 m., July 15 (No. 535).

Bursa bursa-pastoris (L.) Weber in Wigg. Prim. Fl. Holst. 41 (1780); *Thlaspi bursapastoris* L. Sp. Pl. ii, 647 (1753).

The common form was collected in yards at Custer, altitude 1,625 m., June 4 (No. 536).

The so-called variety *integrifolia*, that is, the form with entire leaves, was collected on a hillside near Central City, altitude 1,700 m., July 3 (No. 537).

A delicate form was found with finely pinnatifid leaves; the divisions oblong, sometimes sinnately toothed; pods (immature) broadly oval, sometimes truncate, but not at all triangular. The general appearance reminds one somewhat of *Teesdalia nudicaulis*. Hillside, south of Custer, altitude 1,625 m., May 28 (No. 538).

Lepidium incisum Roth, Neue Beitr. i, 221 (1802).

Rare in the Black Hills: Hot Springs, altitude 1,650 m., June 16 (No. 539).

CAPPARIDACEÆ.

Jacksonia trachysperma (Torr. & Gr.) Greene. Pittonia, ii, 175 (1890); *Polanisia trachysperma* Torr. & Gr. Fl. i, 669 (1840).

Draws among the foothills: Hot Springs, altitude 1,100 m., August 2 (No. 510).

Cleome serrulata Pursh, Fl. ii, 441 (1814).

Draws among the foothills: Hot Springs, altitude 1,100 m., August 2 (No. 541).

CISTACEÆ.

Helianthemum majus (L.) B. S. P. Prel. Cat. N. Y. 6 (1888); *Lechea major* L. Sp. Pl. i, 90 (1753); *Helianthemum canadense walkera* Evans, Bot. Gaz. xv, 211 (1890).

The only locality hitherto recorded for this form is the original one in Douglas County, Colorado. Roadside, east of Custer, altitude 1,600 m., July 22 (No. 542).

VIOLACEÆ.

Viola pedatifida Don, Hist. Dichl. Pl. i. 320 (1831).
Valley: Custer, altitude 1,650 m., June 4 (No. 543).

Viola obliqua Hill, Hort. Kew. 316, t. 12 (1769); *Viola palmata obliqua* (Hill) Hitchcock, Trans. St. Louis Acad. v, 487 (1891); *V. cucullata* Ait. Hort. Kew. iii, 288 (1789).
Low grounds: Ruby Gulch, near Custer, altitude 1,650 m., May 31 (No. 544).

Viola palustris L. Sp. Pl. ii, 934 (1753).
Only four specimens in fruit were collected. These have larger and thinner leaves than usual, resembling somewhat those of the Alaskan *V. langsdorfii*. Sylvan Lake, 6 miles northeast of Custer and 4 miles from Harneys Peak, altitude 2,000 m., July 20 (No. 545).

Viola blanda Willd., Hort. Berol. t. 24 (1806).
Among moss, in a canyon below Sylvan Lake, altitude 1,900 m., June 9 (No. 546).

Viola canina adunca (Smith) Gray, Proc. Amer. Acad. viii, 377 (1872); *Viola adunca* Smith, Rees's Cycl. No. 63 (1817).
My specimens lack the dark-brown spots attributed to this variety in Hook. Fl. Bor. Amer. i, 79. Borders of rich woods, early in the spring. South of Custer, altitude 1,650 m., May 30 (No. 547).

Viola canina oxyceras (?) Wats. Bot. Cal. i, 56 (1876).
I have not seen any specimens of this variety, but as it is the only one described with acute spur, I place this plant with it.
Rare: near Lead City, altitude 1,500 m., July 6 (No. 548).

Viola canadensis L. Sp. Pl. ii, 936 (1753).
Common: Little Elk Canyon, altitude 1,100 m., June 29; Elk Canyon, altitude 1,200 m., June 29; Rochford, altitude 1,600 m., July 11 (No. 549).

Viola nuttallii Pursh, Fl. i, 174 (1814).
Prairies and rich valleys: Custer, altitude 1,650 m., June 3; Hot Springs, altitude 1,050 m., June 12 (No. 550).

Viola pubescens Ait. Hort. Kew. iii, 290 (1789).
Rare: Elk Canyon, altitude 1,200 m., June 29 (No. 551).

Viola scabriuscula (Torr. & Gr.) Schwein.; Torr. & Gr. Fl. i, 142 (1838), as synonym; *Viola pubescens scabriuscula* Torr. & Gr. loc. cit.
This plant has nearly always one or more radical leaves at the time of blooming, while *V. pubescens* seldom has any.
Elk Canyon, altitude 1,100 m., June 29 (No. 552).

POLYGALACEÆ.

Polygala senega latifolia Torr. & Gr. Fl. i. 131 (1838).
It is not uncommon in the northern part of the Black Hills. Elk Canyon, altitude 1,300 m., June 30; south of Lead City, altitude 1,600 m., July 6; near Bull Springs in the Limestone District, altitude 1,900 m., July 26 (No. 553).

Polygala alba Nutt. Gen. ii, 87 (1818).
Hills below Hot Springs, altitude 1,000 m., June 17; 15 miles east of Custer, altitude 1,400 m., July 22 (No. 554).

Polygala verticillata L. Sp. Pl. ii, 706 (1753).
Fifteen miles east of Custer, on French Creek, altitude 1,100 m., July 22 (No. 555).

CARYOPHYLLACEÆ.

Saponaria vaccaria L. Sp. Pl. i, 409 (1753).
On the railroad embankment south of Minnekahta, altitude 1,270 m., August 4 (No. 556).

Silene antirrhina L. Sp. Pl. i, 419 (1753).

488

Very variable. One specimen is about 1 m. high and has broad leaves and minutely pubescent stem; some are only about 1 dm. high and wholly glabrous.

Hermosa, altitude 1,050 m., June 22; Lead City, altitude 1,700 m., July 6; Custer, altitude 1,650 m., August 1 (No. 557).

Lychnis drummondii (Hook.) Wats. Bot. King Surv. 37 (1871); *Silene drummondii* Hook. Fl. Bor. Amer. i, 89 (1830).

It was collected only in fruit. The leaves are unusually narrow and the plant strict. Custer, altitude 1,650 m., August 16 (No. 558).

Cerastium longipedunculatum Muhl. Cat. 16 (1813).

This is very variable. Some are 3.5 to 4.5 dm. high, with large leaves 3 to 5 cm. long and 8 to 12 mm. wide, oblong, oval-lanceolate or spatulate. Rapid City, altitude 1,000 m., June 25; south of Lead City, altitude 1,500 m., July 6; Rochford, altitude 1,600 m., July 11 (No. 559).

The more common form, about 2 to 3 cm. high, with leaves about 25 mm. long and 8 mm. wide, oval or broadly oblong, was collected near Lead City, altitude 1,500 m. July 6; Custer, altitude 1,650 m., June 4 (No. 560).

Cerastium brachypodum (Engelm.) Robinson, Mem. Torr. Club, v, 150 (1894); *Cerastium nutans brachypodum* Engelm.; Gray, Man. ed. 5, 91 (1867).

My specimens are small and approach the variety *compactum*,¹ to which some of them could be referred. Hermosa, on poor soil, altitude 1,050 m., June 22 (No. 561).

Cerastium arvense L. Sp. Pl. i, 438 (1753).

Rarer than the next: Elk Canyon, altitude 1,200 m., June 29; Lead City, altitude 1,700 m., July 6 (No. 562).

Cerastium arvense oblongifolium (Torr.) Britton & Hollick, Bull. Torr. Club, xiv, 17 (1887); *Cerastium oblongifolium* Torr. Fl. U. S. 460 (1824).

Custer, altitude 1,700 m., May 31; Hot Springs, altitude 1,100 m., June 18; Elk Canyon, altitude 1,300 m., June 29 (No. 563).

Alsine longifolia (Muhl.) Britton, Mem. Torr. Club, v, 150 (1894); *Stellaria longifolia* Muhl.; Willd. Enum. 479 (1809).

Grassy places, near water: below Terrys Peak, altitude 1,800 m., July 8; near Sylvan Lake, altitude 2,000 m., July 18 (No. 564).

Arenaria hookeri Nutt.; Torr. & Gray. Fl. i, 178 (1838).

Battle Mountain, east of Hot Springs, altitude 1,150 m., June 18 (No. 565).

Arenaria lateriflora L. Sp. Pl. i, 423 (1753).

Shady woods: Elk Canyon, altitude 1,200 m., June 29; Sylvan Lake, altitude 2,000 m., July 18 (No. 568).

Arenaria stricta Mx. Fl. i, 274 (1803).

Differs from the common form in the length of the petals, which scarcely exceed the acute but not pointed sepals. The leaves are also somewhat broader and more flaccid. Sandy soil: Elk Canyon, altitude 1,200 m., June 29; Little Elk Canyon, altitude 1,100 m., June 28 (No. 566).

Arenaria verna hirta (Wormsk.) Wats. Bot. King Surv. 41 (1871); *Arenaria hirta* Wormsk. Fl. Dan. x. 1646 (1819).

Glandular-puberulent; the upper leaves lanceolate, short, strongly 3-nerved. Shady place among rocks, below Sylvan Lake, altitude 1,900 m., July 18 (No. 567).

Paronychia jamesii Torr. & Gr. Fl. i, 170 (1838).

Dry hills: Hot Springs, altitude 1,100 m., June 13; Hermosa, altitude 1,100 m., June 23 (No. 959).

PORTULACACEÆ.

Talinum parviflorum Nutt.; Torr. & Gr. Fl. i, 197 (1838).

Among rocks, on the south side of Buckhorn Mountain, north of Custer, altitude 1,700 m., July 15 (No. 569).

¹ Robinson, Proc. Amer. Acad. xxix, 278 (1894).

Claytonia perfoliata amplectens Greene, Fl. Fran. 179 (1891).
It is smaller than the typical *C. perfoliata*, has smaller flowers and shorter pedicels; the involucral leaves united only on one side. It has been collected by Dr. Vasey, in the State of Washington, and by C. V. Piper, near Seattle, same State.
Hills, near Whitewood, altitude 1,200 m., July 7 (No. 570).

HYPERICACEÆ.

Hypericum canadense L. Sp. Pl. ii, 785 (1753).
Low grounds, north of Custer, altitude 1,700 m., August 20 (No. 572).

MALVACEÆ.

Malveopsis coccinea (Nutt.) Kuntze, Rev. Gen. Pl. 72 (1891); *Malva coccinea* Nutt. Fraser's Cat. (1813).
Hillside, above Hot Springs, altitude 1,100 m., June 11 (No. 580).

LINACEÆ.

Linum lewisii Pursh, Fl. i, 210 (1814).
Hillsides: Hot Springs, altitude 1,100 m., June 11; Elk Canyon, altitude 1,200 m., June 29; Rochford, altitude 1,700 m., July 12; Custer, altitude 1,700 m., July 15 (No. 581).

Linum rigidum Pursh, Fl. i, 210 (1814).
Rare: Hot Springs, altitude 1,100 m., June 15 (No. 582).

GERANIACEÆ.

Geranium richardsonii Fisch. & Mey. Ind. Sem. Petrop. iv, 37 (1837).
The most common species in the Black Hills. The flowers are nearly always white or light pinkish with purple veins. Valleys: Elk Canyon, altitude 1,300 m., June 29; Rochford, altitude 1,600 m., July 11; Sylvan Lake, altitude 2,000 m., July 21 (No. 583).

Geranium viscosissimum Fisch. & Mey. Ind. Sem. Petrop. xi, Suppl. 18 (1813); *Geranium incisum* Nutt.: Torr. & Gray, Fl. i, 206 (1838), not Andrews, Bot. Rep. 67 (1814).
Rochford, altitude 1,600 m., July 11; Bull Springs, altitude 1,900 m., July 27 (No. 584).

Geranium carolinianum L. Sp. Pl. ii, 682 (1753).
Elk Canyon, altitude 1,200 m., June 29; Whitewood, altitude 1,100 m., July 7; Custer, altitude 1,650 m., July 15 (No. 585).

Oxalis stricta L. Sp. Pl. i, 435 (1753).
Shady places among the foothills: Rapid Creek, altitude 1,100 m., June 25; Elk Canyon, altitude 1,200 m., June 29 (No. 586).

CELASTRACEÆ.

Celastrus scandens L. Sp. Pl. i, 196 (1753).
Erect, 1 to 1.5 m. high; nowhere found climbing. Rapid City, altitude 1,050 m., June 25; Little Elk Canyon, altitude 1,100 m., June 28; Lead City, altitude 1,600 m., July 6 (No. 587).

RHAMNACEÆ.

Ceanothus velutinus Dougl.; Hook. Fl. Bor. Amer. i, 125 (1830).
Not uncommon in the hills around Lead City, altitude 1,700 m., July 4 (No. 588).

Ceanothus ovatus Desf. Hist. Arb. ii, 381 (1809).
The common peduncles are in most cases elongated, and the leaves are thinner

than in the Nebraska specimens. Little Elk Canyon, altitude 1,100 m., June 28; Elk Canyon, altitude 1,200 m.. June 29; south of Lead City, altitude 1,700 m., July 6 (No. 589).

Ceanothus fendleri Gray. Pl. Fendl. 29 (1849).

This is common in the Limestone District, west of Custer. Bull Springs, altitude 2,000 m., July 26 (No. 590).

VITACEÆ.

Vitis vulpina L. Sp. Pl i, 203 (1753).

Hermosa, altitude 1,050 m.. June 24; Little Elk, altitude 1,100 m., June 28; Elk Canyon, altitude 1,200 m., June 29 (No. 591).

Parthenocissus quinquefolia (L.) Planch. in DC. Monogr. Phan. v, pt. 2. 488 (1887); *Hedera quinquefolia* L. Sp. Pl. i, 202 (1753).

Elk Canyon, altitude 1,200 m., June 29 (No. 592).

ACERACEÆ.

Acer negundo L. Sp. Pl. ii, 1056 (1753).

In fruit: Hot Springs, altitude 1,050 m., June 18 (No. 593).

ANACARDIACEÆ.

Rhus radicans toxicodendron (L.) Pers. Syn. Pl. i, 325 (1805); *Rhus toxicodendron* L. Sp. Pl. i, 266 (1753).

For remarks on this form of the poison ivy, see page 152 of this volume.
In the foothills: Hermosa, altitude 1,100 m., June 28 (No. 594).

Rhus trilobata Nutt. ; Torr. & Gr., Fl. i, 219 (1838).

I found stems of this shrub 2.5 to 3 m. high and 7 to 8 cm. in diameter, and with leaflets 3 to 4 cm. long. Hot Springs, altitude 1,100 m., June 11 and August 8 (No. 595).

PAPILIONACEÆ.

Thermopsis rhombifolia (Pursh) Richards. Bot. App. 737 (1823); *Cytisus rhombifolia* Pursh, Fl. Suppl. 741 (1811).

Common : Custer, altitude 1,700 m., June 23 (No. 596).

Lupinus sericeus Pursh, Fl. ii. 468 (1814), var.

This is *L. ornatus* Dougl., var., of Newton & Jenney's Report. [1]
The Black Hills and Wyoming specimens have the flowers dark blue and the calyx little gibbous. I think, however, they belong to this species rather than to *L. leucophyllus*, with which in some cases they have been placed. Elk Canyon, altitude 1,200 m., June 29; near Bull Springs in the Limestone District, altitude 1,900 m., July 27 (No. 597).

Lupinus parviflorus Nutt.; Hook & Arn. Bot. Beech. 336 (1840).

Common: Little Elk Canyon, altitude 1,100 m., June 28; Elk Canyon, altitude 1,200 m., June 29; Lead City, altitude 1,700 m., July 4; Bull Springs, altitude 1,900 m., July 28 (No. 598).

Lupinus pusillus Pursh, Fl. ii, 468 (1814).

Hill near Fall River Falls, altitude 1,000 m.. June 17 (No. 599).

Trifolium pratense L. Sp. Pl. i, 768 (1753).

Escaped in meadows: Buffalo Gap, altitude 975 m., June 21; Hot Springs, altitude 1,050 m., August 2 (No. 600).

Trifolium repens L. Sp. Pl. ii, 767 (1753).

Meadows: Custer, altitude 1,650 m., August 1 (No. 601).

Lotus americanus (Nutt.) Bisch. Litt. Ber. Linnæa, xiv, 132 (1840); *Trigonella americana* Nutt. Gen. ii, 120 (1818).

Hills, near Whitewood, altitude 1,150 m., July 7 (No. 602).

[1] Geol. Surv. Black Hills, 532 (1880).

Psoralea tenuiflora Pursh, Fl. ii, 475 (1814).

In the foothills, 15 miles east of Custer, altitude 1,400 m.. July 23 (No. 603).

Psoralea argophylla Pursh, Fl. ii, 475 (1814).

Rochford, altitude 1,600 m., July 12; Custer, altitude 1,650 m.. July 15: Hot Springs, altitude 1,100 m., August 8 (No. 604).

Psoralea cuspidata Pursh, Fl. ii. 741 (1814).

Among the foothills: Fall River Falls, altitude 1,050 m., August 10 (No. 605).

Psoralea esculenta Pursh, Fl. ii, 475 (1814).

The specimens from Lead City are low, 7 to 10 cm. high, with obovate leaflets and whitish flowers. Hot Springs. altitude 1,100 m., June 18; Hermosa, altitude 1,050 m., June 23; Lead City, altitude 1,700 m., July 4; Rochford, altitude 1,700 m., July 12 (No. 606).

Parosela enneandra (Nutt.) Britton, Mem. Torr. Club, v, 196 (1894); *Dalea enneandra* Nutt. Fraser's Cat. 1813.

Among the foothills: Hot Springs, altitude 1,100 m., August 2 (No. 607).

Parosela aurea (Pursh) Britton, Mem. Torr. Club, v, 196 (1894): *Dalea aurea* Nutt.; Pursh, Fl. ii, 740 (1814).

Among the foothills: Hot Springs, altitude 1,150 m., August 2 (No. 608).

Amorpha canescens Pursh, Fl. ii, 467 (1814).

Hills, 15 miles east of Custer, altitude 1,400 m., July 22; Hot Springs, altitude 1,100 m., August 2 (No. 609).

Amorpha fruticosa L. Sp. Pl. ii, 713 (1753).

This was seen growing along French Creek and Fall River, but no specimens were secured.

Kuhnistera purpurea (Vent.) MacMillan, Metasp. Minn. Val. 329 (1892); *Dalea purpurea* Vent. Hort. Cels. t. 40 (1800).

Variable. Some of the specimens are pubescent on the stem, but most of them are glabrous and have short spikes with the calyx woolly, rather than silky. In the Limestone District, altitude 1,900 m., July 27; Hot Springs, altitude 1,100 m., August 2 (No. 610). Some of the specimens from the latter place have white flowers (No. 611).

Kuhnistera candida occidentalis Rydberg, Contr. Nat. Herb. iii, 154 (1895).

In the specimens of this collection the bracts are shorter than the calyx, in which respect they approach the variety *multiflora*.

Hot Springs, altitude 1,100 m., August 2 (No. 612).

Astragalus crassicarpus Nutt. Fraser's Cat. 1813.

Nearly all my specimens from the Black Hills have large (16 to 20 mm. long), ochroleucous flowers. purplish only at the tip; but without doubt they all belong to *A. crassicarpus*. Custer, altitude 1,650 m., May 31 to June 1 (No. 613).

Astragalus sp.

The specimens are without pods, hence can not well be determined. The plant may be a form of the preceding, which it resembles, though more upright, ascending; racemes elongated (1.5 dm. long), flowers distant on upright pedicels, bracts about 4 mm. long, calyx appressed-hairy with dark hairs; corolla dark purple. Only one plant found, that in full bloom on Battle Mountain, east of the Hot Springs, altitude 1,200 m., June 18 (No. 614).

Astragalus plattensis Nutt.; Torr. & Gr. Fl. i, 332 (1838).

Not uncommon in the southern part of the Black Hills. Hot Springs, altitude 1,100 m., June 13, August 3 (No. 615).

Astragalus carolinianus L. Sp. Pl. ii, 757 (1853).

Rochford, altitude 1.700 m., July 12; French Creek, 15 miles below Custer, altitude 1,500 m., July 23 (No. 616).

Astragalus adsurgens Pall. Astrag. 10, t. 31 (1800); *Astragalus larmanni* Pall. (?), Nutt. & authors, not Jacq. Lately the name *A. larmanni* Jacq.,[1] has been adopted for our American plant. Although there is nothing in the original description that really disagrees with our plant, yet the plate accompanying it shows that Jacquin's *A. larmanni* was of a different habit. The stem is very slender and decumbent, the leaflets narrower and smaller, the heads, or rather spikes, much longer and narrower. The pods seem to be like those of *A. adsurgens*, but are more truncate at the apex and have the style abruptly turned dorsally, making a right angle with the pod.[2] In *A. adsurgens* the pod acuminates into a nearly central style, which is somewhat twisted and curved dorsally, but does not make a right angle. In the Columbia College Herbarium there is a specimen of *A. larmanni* collected by A. Regel in Turkestan, which perfectly agrees with the figure in Hortus Vindobonensis. The flowers of this, although of about the length of those of *A. adsurgens*, are much narrower, the calyx less than one-half the length of the claws of the petals, or with the teeth about two-thirds their length, while in *A. adsurgens* the calyx with the teeth nearly equals the claws.

Common in the region. Some specimens from Hot Springs have a more decumbent stem and brighter blue flowers. Hot Springs, altitude 1,100 m., June 8; Hermosa, altitude 1,050 m., June 22; Custer, altitude 1,700 m., July 15 (No. 617).

Astragalus hypoglottis L. Mant. ii, 274 (1771).
Not uncommon among the foothills: Hot Springs, altitude 1,100 m., June 11; Hermosa, altitude 1,050 m., June 22 (No. 618).

Astragalus drummondii Dougl.; Hook. Fl. Bor. Amer. i, 153 (1834).
Hills near Hot Springs, altitude 1,100 m., June 18 (No. 619).

Astragalus racemosus Pursh, Fl. ii, 740 (1814).
The corolla is ochroleucous rather than white, and the angles of the pods are blunter than in Nebraska specimens collected in 1891. Near Fall River Falls, altitude 1,000 m., June 17 (No. 620).

Astragalus gracilis Nutt. Gen. ii, 100 (1818).
Hot Springs, altitude 1,100 m., June 15 (No. 573).

Astragalus microlobus Gray, Proc. Amer. Acad. vi, 203 (1861).
Hot Springs, altitude 1,100 m., June 15 (No. 621).

Astragalus aboriginum Richards. Bot. App. 746 (1823).
The fruit is membranaceous, long-stipitate, strictly 1-celled, but the dorsal suture a little inflexed, straight, the ventral one curved. It was collected in fruit only, north of Deadwood, altitude 1,500 m., July 5 (No. 622).

Astragalus aboriginum glabriusculus (Hook.): *Phaca glabriuscula* Hook. Fl. Bor. Amer. i, 144 (1830).
This has generally been regarded as a distinct species, but even Hooker says, in the original description, that it may be a variety of the preceding. The only difference I can find is that the plant is smoother and the pod a little more curved. The flowers are ochroleucous, the keel tipped with purple. Custer, altitude 1,700 m., June 1; Rochford, altitude 1,700 m., July 12; Limestone District, altitude 2,000 m., July 26 (No. 623).

Astragalus alpinus L. Sp. Pl. ii, 760 (1753).

---

[1] Hort. Vind. iii, 22 (1776).

[2] There is a species from Japan, much larger but with the same pod characters, in the National Herbarium. This was identified by Bunge, the well-known authority on Old World Astragali, as *A. adsurgens*, but is evidently distinct. It differs in the style, and in its more slender, less distinctly striate stem, its looser heads on peduncles which are neither strict nor longer than the leaves. Notwithstanding Bunge's determination the plant can not be *A. adsurgens*, this name belonging to our species, as is plainly shown by the original plates.

In woods: Custer, altitude 1,700 m., June 3; Little Elk Canyon, altitude 1,200 m., June 28; Lead City, altitude 1,600 m., July 6; Rochford, altitude 1,700 m., July 11 (No. 624).

Astragalus lotiflorus Hook. Fl. Bor. Amer. i, 152 (1834).

In nearly all the specimens of my collection the flowers are in short, capitate racemes, but at least in some a part of the flowers are sessile, the plant thus approaching the forma *brachypus* Gray. Mr. E. P. Sheldon has raised the latter to specific rank, but it can scarcely be regarded even as a variety, and Dr. Gray seems to me to have disposed of it correctly.

Hot Springs, altitude 1,100 m., June 16 (No. 625).

Astragalus missouriensis Nutt. Gen. ii, 99 (1818).

The specimens from the Black Hills are, like those from Nebraska, greener than the species generally is farther south. Hot Springs, altitude 1.100 m., June 13 (No. 626). A few specimens collected at Hermosa, altitude 1,000 m., June 23, differ considerably from the common form. The stem is not cespitose, but ascending, less hairy; leaflets larger, obovate: flowers racemose or subcapitate on a long peduncle. In general appearance, color of the flowers, etc., they resemble *A. adsurgens*, but the heads are more lax, the leaflets broader, and the plant more hairy. The pubescence, although sparser, is that of *A. missouriensis*, so also the pod. It may be a hybrid between the two, which were found growing together (No. 627).

Astragalus frigidus americanus (Hook.) Wats. Ind. 193 (1878); *Phaca frigida americana* Hook. Fl. Bor. Amer. i, 140 (1830).

The flowers in my specimens are light ochroleucous. Low ground in shade: Rochford, altitude 1,600 m., July 12 (No. 628).

Astragalus bisulcatus (Hook.) Gray, Pac. R. Rep. xii, bk. ii, pt. ii, 42 (1860); *Phaca bisulcata* Hook. Fl. Bor. Amer. i, 145 (1834).

Plains among the foothills: Hermosa, altitude 1.000 m., June 23 (No. 629).

Astragalus flexuosus Dougl.; Hook. Fl. Bor. Amer. i. 141 (1834), as synonym; *Phaca flexuosa* Hook. loc. cit.

The specimens are large and decumbent. Rochford, altitude 1.700 m., July 12 (No. 630).

Astragalus convallarius Greene, Erythea, i. 207 (1893).

Only a few slender specimens were collected. The flowers are unusually small, ochroleucous. Bull Springs, altitude 2.000 m., July 29 (No. 631).

Astragalus tenellus Pursh, Fl. ii, 473 (1814).

Near Bull Springs, altitude 1,900 m., July 27 (No. 632).

Astragalus spatulatus Sheld. Geol. & Nat. Hist. Surv. Minn. Bull. 9, pt. i, 22 (1891). Probably *A. simplicifolius* Gray, of Newton & Jenney's Report. [1]

Hot Springs, altitude 1,100 m., June 13; Bull Springs, altitude 1,900 m., July 27 (No. 633). Near the latter place some specimens were found with 3 and 2-foliate leaves, showing that the common form has developed by reduction. July 28 (No. 634).

Astragalus gilviflorus Sheldon. Geol. & Nat. Hist. Surv. Minn. Bull. 9, pt. i, 21 (1891).

Foothills: Hot Springs, altitude 1,100 m., June 18 (No. 635).

Spiesia viscida (Nutt.) Kuntze, Rev. Gen. Pl. i, 207 (1891); *Oxytropis viscida* Nutt.; Torr. & Gr. Fl. i, 341 (1838); *Oxytropis monticola* Gray, Proc. Amer. Acad. xv, 6 (1885).

My specimens differ from all except one of those named *O. viscida* in the National Herbarium in being more silky and having larger, yellowish flowers in an elongated spike. The one excepted was collected by Wm. C. Cusick in Oregon, which is nearer to the type locality of Nuttall's plant than any of the localities represented by the other specimens. Nuttall's type locality was, "near the sources of the Oregon"

[1] Geol. Surv. Black Hills, 533 (1880).

(Columbia). In the Gray Herbarium there is a poor specimen of the original *O. viscida* of Nuttall, and this resembles more my plant, which Dr. Gray would have included in *O. monticola* (Jenney's plants from the Black Hills are included in the latter), rather than in *O. viscida* as understood by him.

Common around Custer, altitude 1,650 to 1,700 m., June 5 (No. 636).

Spiesia lambertii (Pursh) Kuntze, Rev. Gen. Pl. i, 207 (1891); *Oxytropis lambertii* Pursh, Fl. ii, 740 (1814).

Hot Springs, altitude 1,100 m., June 18; Rochford, altitude 1,700 m., July 11 (No. 638).

Spiesia lambertii sericea (Nutt.) Rydberg, Bot. Surv. Nebr. iii, 32 (1894); *Oxytropis sericea* Nutt.; Torr. & Gr. Fl. i, 339 (1838).

All my specimens have yellow flowers. In a few the calyx is somewhat viscid, and these can scarcely be distinguished from the preceding species except by the size. The bracts are narrower than in the blue-flowered forms I have seen.

Hot Springs, altitude 1,100 m., June 16 (No. 637).

Glycyrrhiza lepidota Pursh, Fl. ii, 480 (1814).

On the French Creek, 15 miles below Custer, altitude 1,400, July 22 (No. 639).

Hedysarum americanum Britton, Mem. Torr. Club. v, 201 (1891); *Hedysarum alpinum americanum* Mx. Fl. ii, 74 (1803); *H. boreale* Nutt. Gen. ii, 110 (1818).

In Coulter's Manual the stamens are given as diadelphous (5 and 1), in Gray's Manual as diadelphous (9 and 1) in the key, but as 5 and 1 in the description of the genus. In all flowers investigated, they were 9 and 1, but the united stamens were of two different lengths, every second one being shorter.

Hills: Rochford, altitude 1,700 m., July 12 (No. 610).

Vicia americana Muhl.; Willd. Sp. Pl. iii, 1096 (1801).

Common: Little Elk Canyon, altitude 1,100 m., June 28; Elk Canyon, altitude 1,200 m., June 29; Lead City, altitude 1,600 m., July 6 (No. 641).

Vicia americana linearis (Nutt.) Wats. Proc. Amer. Acad. xi, 134 (1876); *Lathyrus linearis* Nutt.; Torr. & Gr. Fl. i, 276 (1838).

V. americana and this plant grow together sparsely everywhere in Nebraska. I have not seen any intermediate forms, but they both grade into the variety *truncata*. Hot Springs, altitude 1,050 m., June 16 (No. 643).

Vicia americana truncata (Nutt.) Brewer, Bot. Cal. i, 158 (1856); *Vicia truncata* Nutt.; Torr. & Gr. Fl. i, 270 (1838).

Always near water, where *V. americana* and *V. linearis* are found on the drier land. Hot Springs, altitude 1,050 m., June 13 (No. 642).

Lathyrus ochroleucus Hook. Fl. Bor. Amer. i., 159 (1833).

Common and very luxurious in the Black Hills. It is regarded as a very good fodder plant, and may be of economic value.

Rapid Creek, altitude 1,000 m., June 25; Elk Canyon, altitude 1,200 m., June 29 (No. 644).

ROSACEÆ.

Prunus americana Marsh. Arb. Amer. 111 (1785).

Draws among the foothills: Minnekahta, altitude 1.275 m., August 4 (No. 645).

Prunus besseyi Bailey, Bull. Cornell Agr. Exp. Sta. 70, 261 (1894).

In the foothills: Hermosa, altitude 1,050 m., June 23; Minnekahta, altitude 1,300 m., August 4 (No. 646).

Prunus pennsylvanica L. f. Suppl. 252 (1781).

In the Black Hills it never becomes a large tree. The largest specimens I saw were less than 7 m. high. On the hills around Lead City, it is only a small shrub, not much taller than the preceding, and has generally folded leaves.

Woods: Custer, altitude 1,700 to 1,800 m.; Lead City, altitude 1,700 m., July 3 and 6 (No. 617).

Prunus virginiana L. Sp. Pl. i, 473 (1753).

Wholly glabrous; leaves dull, ovate with acuminate base, thin, sharply serrate. In the Black Hills only a shrub or a low tree, at most 6 cm. in diameter. In damp, shaded canyons: Hot Springs, altitude 1,100 m., June 13; Sylvan Lake, altitude 2,000 m., July 18 (No. 648).

Prunus demissa (Nutt.) Walp. Rep. ii, 10 (1843); *Cerasus demissa* Nutt.; Torr. & Gr. Fl. i, 411 (1840).

I include under this two forms. One is the common chokecherry of the western plains. It is generally glabrous; leaves oval with truncate or even cuneate base, thick, pale beneath. It differs from *P. virginiana* in the much thicker leaves and sweeter fruit. Wherever I have seen it, it is a small tree—that is, it has one principal stem, with a rounded top. It has been named *P. demissa*, although I doubt whether it is identical with the original. Rochford, altitude 1,600 m., July 12 (No. 649).

The other form has the young shoots, peduncles, and lower surface of the leaves pubescent, and even a little viscid, glabrate in age; leaves thick, shining above, paler beneath, elliptical or broadly oval, abruptly pointed or obtuse; base truncate or slightly cordate, or sometimes somewhat cuneate; flowers larger than in the preceding two. A low shrub, generally a few meters high. The largest stem I saw was about 5 cm. in diameter, with heartwood fully as dark as in *P. serotina*. It was growing in the same canyon as *P. virginiana*, from which it was easily distinguished. Hot Springs, altitude 1,100 m., June 13 and August 8 (No. 650).

Spiræa lucida Dougl.; Hook. Fl. Bor. Amer. i, 172 (1833), as synonym; *S. betulifolia* Hook. loc. cit., not Pall.

Professor Greene[1] has separated the American species from the Asiatic.

Rapid City, altitude 1,000 m., June 25; Little Elk, altitude 1,100 m., June 28; Lead City, altitude 1,600 m., July 6 (No. 651).

Luetkea cæspitosa (Nutt.) Kuntze, Rev. Gen. Pl. i, 217 (1891); *Spiraa cæspitosa* Nutt.; Torr. & Gr. Fl. i, 418 (1840).

Neither flowers nor fruit were found. The plant was growing on the hills around Little Elk, altitude 1,300 m., June 28 (No. 652).

Opulaster opulifolius (L.) Kuntze, Rev. Gen. Pl. ii, 949 (1891), var.; *Spiraa opulifolia* L. Sp. Pl. i, 489 (1753).

My specimens from northern Nebraska and those I have seen from Colorado differ from the *Opulaster opulifolius* of the eastern United States in having pubescent ovaries. The pubescence remains, partly at least, until maturity, while in the eastern form the fruits are smooth and shining. The ovaries are generally only three, and the leaves smaller and more rounded in outline. This form seems to connect this species and the following. Seeds obliquely pear-shaped, shining, carinate on one side.

In the lower parts of the Hills near water. Rapid Creek, altitude 1,100 m., June 25; French Creek, 10 miles below Custer, altitude 1,500 m., June 22; Hot Springs, altitude 1,050 m., August 2 (No. 653).

Opulaster monogyna[2] (Torr.) Kuntze, Rev. Gen. Pl. ii, 949 (1891); *Spiraa monogyna* Torr. Ann. Lyc. N. Y. ii, 1892 (1827).

Very shrubby, 3 to 6 dm. high; leaves small, about 2.5 cm. long, round in outline, deeply 3 to 5 cleft, teeth sharper than in the preceding; flowers half the size, ovaries mostly 2, very woolly. The leaves are perfectly smooth, in this point disagreeing with the description of *Neillia torreyi*, but otherwise agreeing with the

[1] Pittonia, ii, 221.

[2] Thus named by Professor Greene. I have seen specimens of an Opulaster, collected by Dr. Sandberg in Idaho, which fits the description of *Spiraa monogyna* Torr., loc. cit., even as to the number of the carpels. This differs as much from my specimens as does *Opulaster opulifolia*.

type specimens. It also disagrees in several points with the description of Professor Greene, especially in the size of the flowers. Perhaps all the Opulasters of North America are but one very variable species.

Hills near Harneys Peak, altitude 2,100 m., July 21, August 17 (No. 651).

Rubus parviflorus Nutt. Gen. i, 308 (1818).

Not uncommon: Elk Canyon, altitude 1,300 m., June 29; Lead City, altitude 1,600 m., July 1 (No. 655).

Rubus americanus (Pers.) Britton. Mem. Torr. Club. v. 185 (1891); *Rubus saxatilis americanus* Pers. Syn. ii, 52 (1807).

Canyon north of Runkels, altitude 1.300 m., June 30; Lead City, altitude 1,600 m., July 4 (No. 656).

Rubus strigosus Mx. Fl. i, 297 (1803).

Little Elk Canyon, altitude 1.200 m., June 28; Elk Canyon, altitude 1,300 m., June 29; Rochford, altitude 1,700 m., July 12 (No. 657).

Cercocarpus parvifolius Nutt.; Hook. & Arn. Bot. Beech. Suppl. 337 (1841).

Very rare: Hot Springs, altitude 1,100 m., June 13 (No. 658).

Geum strictum Ait. Hort. Kew. ii, 217 (1789).

Lead City, altitude 1,600 m., July 6 (No. 659).

Geum macrophyllum Willd. Enum. i, 557 (1809).

I think this would better be regarded as a variety of the preceding. Rapid City, altitude 1,050 m., June 25; Rochford, altitude 1,600 m., July 11 (No. 1206).

Geum ciliatum Pursh. Fl. i, 352 (1814).

This name precedes *G. triflorum* in Pursh's Flora.[1] Common: Custer, altitude 1,700 m., May 28; Elk Canyon, altitude 1,200 m., June 29; Lead City, altitude 1,600 m., July 4 (No. 660.)

Fragaria virginiana Duchesne, Hist. Nat. Frais. 204 (1766).

A low, small-leafed form, collected early in the spring. Custer, altitude 1,650 m., June 4 (No. 661).

Fragaria vesca americana Porter, Bull. Torr. Club, xvii, 15 (1890)

Custer, altitude 1,700 m., June 6; Hot Springs, altitude 1,100 m., June 14; Rapid City, altitude 1,000 m., June 25; in fruit, Custer. August 20 (No. 662).

A slender form with very thin, cuneate, narrow leaves, 3 times as long as broad, toothed towards the apex, the peduncles slender, about the length of the leaves, was collected near Sylvan Lake, altitude 1,900 m., July 18 (No. 663).

Potentilla arguta Pursh, Fl. ii, 736 (1814).

The flowers appear to be always white, but they turn yellow in drying. Hills: Hermosa, altitude 1,100 m., June 22; Lead City, altitude 1,700 m., July 6 (No. 664).

Potentilla glandulosa Lindl. Bot. Reg. xix, t. 1583 (1833).

In this the flowers are light yellow, the cyme more diffuse than usual. Hills: Rochford, altitude 1,700 m., July 12 (No. 665).

Potentilla monspeliensis L. Sp. Pl. i, 499 (1753).

This includes *P. norvegica* L. described lower on the same page. Hot Springs, altitude 1,050 m., June 14; Hermosa, June 22; Rapid City, altitude 1,000 m., June 25; Elk Canyon, altitude 1,200 m., June 29; Rochford, altitude 1,600 m., July 11. The specimens from the latter place are very slender and approach *P. rivalis* (No. 666).

Potentilla pennsylvanica strigosa Pursh, Fl. i, 356 (1814).

Common: Hermosa, altitude 1,050 m., June 23; Rochford, altitude 1,700 m., July 11; Custer, altitude 1,625 m., August 13 (No. 667).

Potentilla hippiana Lehm. Nov. Stirp. Pug. ii, 7 (1830).

Rochford, altitude 1,600 m., July 12; Custer, altitude 1,650 m., July 18 (No. 668). These specimens seem to be typical, agreeing fully with the description and plate in

[1] Pursh, Fl. ii, 736.

Hooker's Flora, except that the branches of the cyme are more upright and the calyx lobes longer. A slender form approaching *P. effusa* was collected at Bull Springs, July 26 (No. 669).

Potentilla hippiana diffusa (Gray) Lehm. Add. Ind. Hort. Hamb. 8 (1849); *P. diffusa* Gray, Pl. Fendl. 11 (1849).
Custer, 1,650 m., August 13 (No. 669½).

Potentilla gracilis Dougl.; Hook. Bot. Mag. lvii, t. 2984 (1830).
This plant has always been regarded as *P. gracilis* Dougl., but there is a specimen so labeled, collected by Mr. Douglas at Puget Sound, in the Columbia College herbarium, and it seems quite different.
Lead City, altitude 1,700 m., July 1; Rochford, altitude 1,600 m., July 12 (No. 670).

Potentilla gracilis fastigiata (Nutt.) Wats. Proc. Amer. Acad. viii, 557 (1873); *P. fastigiata* Nutt.; Torr. & Gr. Fl. i, 440 (1838).
Rochford, altitude 1,550 m., July 11 (No. 671).

Potentilla nivea dissecta Wats. Proc. Amer. Acad. viii, 559 (1873).
This form was included in Dr. Watson's variety, but it seems to have very little relationship to *P. nivea*. It appears to be connected rather with *P. concinna humistrata* and forms of *P. gracilis*. The name should also be changed, as there is an older *P. dissecta* Pursh. It will be left, however, in the present form until its relationship is settled.
Rare: Hot Springs, altitude 1,100 m., June 11 (No. 672).

Potentilla concinna Richards. App. Frankl. Journ. ed. 2, 20 (1823).
Custer, altitude 1,650 m., June 3 (No. 673).

Potentilla concinna humistrata, nom. nov.; *P. concinna humifusa* (Nutt.) Lehm. Rev. Pot. 112 (1856); *P. humifusa* Nutt. Gen. i, 310 (1818), not Willd.; Schlecht. Gesell. Naturf. Freunde Berlin Mag. vii. 289 (1813).
Hills north of Deadwood, altitude 1,500 m., July 5 (No. 673½).

Potentilla fruticosa L. Sp. Pl. i, 495 (1753).
Dry hills and mountain sides: Elk Canyon, altitude 1,300 m., June 29; Rochford, altitude 1,700 m., July 12 (No. 674).

Agrimonia striata Mx. Fl. i, 287 (1803).
Custer, altitude 1,650 m., July 18; Rochford, altitude 1,600 m., July 12 (No. 675).

Rosa engelmanni Wats. Gard. & For. ii, 376 (1889).
Common in the Black Hills: Little Elk Canyon, altitude 1,100 m., June 28; Lead City, July 4; Rochford, July 12; Custer, altitude 1,700 m., August 19 (No. 676).

Rosa woodsii Lindl. Ros. Monogr. 21 (1820).
The sepals are, however, seldom lobed. A character so unstable, should never be used to distinguish the roses.[1] Hot Springs, altitude 1,050 m., July 17; Hermosa, altitude 1,000 m., June 21; Little Elk Canyon, altitude 1,100 m., June 27 (No. 677).

Rosa arkansana Porter, Port. & Coult. Fl. Col. 38 (1874).
I thought at first that this must be a form of *R. humilis*, as the sepals are deciduous, but the leaflets are different; there are no infrastipular spines, and the calyx is not setose. It agrees best with *R. arkansana*, except as to the deciduous sepals. A very low shrub, only 1 to 2 dm. high, growing on a very dry hill near Hermosa, altitude 1,100 m., June 23 (No. 678).

Cratægus macrantha Lodd.; Loud. Arb. & Frut. ed. 2, ii, 819 (1851).
Among the foothills: Hermosa, altitude 1,100 m., June 23; canyon north of Runkels, altitude 1,300 m., June 30 (No. 679).

Amelanchier alnifolia Nutt.; Torr. & Gr. Fl. i, 473 (1840), as synonym; *Aronia alnifolia* Nutt. Gen. i, 306 (1818).

---

[1] The character is not given in the original description (Lindley, on the contrary, states that the sepals are entire) and seems to have been added by Dr. Watson.

Leaves densely white-woolly beneath when young, but wholly glabrous when mature. The leaves of my specimens are unusually thin. Custer, altitude 1,700 m., June 5 and July 15; Hermosa, altitude 1,100 m., June 24; Elk Canyon, altitude 1,300 m., June 29 (No. 680).

Sorbus sambucifolia (Cham. & Schlecht.) Roem. Syn. Mon. iii, 39 (1847); Cham. & Schlecht. Linnæa, ii, 36 (1827).

According to Mr. Runkel, an enterprising lumberman, this is growing in a canyon on the road between Runkel's sawmills and Sturgis. I did not see any specimens in the Black Hills.

SAXIFRAGACEÆ.

Saxifraga cernua L. Sp. Pl. i. 403 (1753).

A few slender specimens in bud, collected above Sylvan Lake, altitude 2,000 m., July 8 (No. 681).

Tellima parviflora Hook. Fl. Bor. Amer. i, 239 (1833).

Rare: Custer, altitude 1,700 m., June 4; west of Lead City, altitude 1,800 m., July 4 (No. 682).

Heuchera hispida Pursh. Fl. i. 188 (1814).

Common: Hot Springs, altitude 1,100 m., June 8; Rochford, altitude 1,700 m., July 12 (No. 683).

Heuchera parvifolia Nutt.; Torr. & Gr. Fl. i, 581 (1840).

Only one specimen secured at Rochford, altitude 1,700 m., July 12 (No. 684).

Parnassia parviflora DC. Prodr. i, 320 (1824).

French Creek, below Custer, altitude 1,600 m., August 1 (No. 685).

Ribes setosum Lindl. Trans. Hort. Soc. vii, 243 (1830).

This agrees with the description, also by Lindley, in the Botanical Register,[1] except that the berries are rarely bristly. Dr. Gray says, in the American Naturalist:[2] "The young berries, either perfectly smooth and naked, or beset with a few bristly prickles." It has been taken for R. oxycanthoides L. (R. hirtellum Mx.), from which it differs in that the leaves are finely pubescent, the calyx cylindrical and longer than the lobes. The bush is generally more spiny and prickly, and the berry sometimes a little bristly, dark purple, and extremely sour even when ripe. It has been found in northwestern Nebraska by Professor Swezey, of the University of Nebraska, who was the first to recognize it as R. setosum Lindl.

Most specimens in herbaria under the name R. setosum are not R. setosum of Lindley, but of Gray,[3] which is a variety of R. lacustre Poir. or a related species. It resembles R. setosum Lindl. somewhat in general habit, but the flowers and fruit are different.

Very common in the Black Hills: Custer, altitude 1,700 to 1,900 m., May 29; in fruit, Minnekahta, altitude 1,300 m., August 5 (No. 686).

Ribes oxycanthoides L. Sp. Pl. i, 201 (1753), var.

I place this plant doubtfully with this species, from which it differs in the longer peduncles and the longer calyx tube. It may also be a form of the preceding, but is nearly without thorns. The leaves are more deeply cleft, with acutish lobes, smooth and shining above, finely and sparingly pubescent beneath, in form resembling somewhat those of R. aureum. The flowers are as in R. setosum, i. e., the calyx cylindrical, a little longer than the narrowly oblong calyx lobes; spines and bristles very rare and small; petioles ciliated by a few fine-fringed bristles. Immature fruit smooth, yellowish; mature fruit not seen. The stem and leaves of R. aureum, with the pubescence and flowers of R. setosum, would fairly represent my plant, which may, perhaps, be a hybrid between the two.

[1] xv, t. 1237 (1829).
[2] x, 271 (1876).
[3] Proc. Amer. Acad. viii, 383 (1872).

Only two small bushes seen in a shady place below Sylvan Lake, altitude 1,900 m., July 18 (No. 687).

Ribes lacustre (Pers.) Poir. Encycl. Suppl. ii, 856 (1811); *Ribes oxyeanthoides lacustre* Pers. Syn. i, 252 (1805).

In shady, damp places; not common: South of Lead City, altitude 1,500 m., July 6; Sylvan Lake, altitude 1,900 m., August 17, (No. 688).

Ribes cereum Dougl. Trans. Hort. Soc. Lond. vii, 512 (1830).

In Coulter's Manual and in the Botany of California this species has been placed in the wrong section, as the calyx is tubular and the foliage glandular. In the Botany of California the form of the calyx is given in the description of the species, but in Coulter's Manual this is ommitted. Specimens collected in flower, therefore, have been named *R. sanguineum variegatum*.

Common: Custer, altitude 1,700 to 1,800, June 11 : 15 miles east of the same place, altitude 1,400 m., July 23 (No. 689).

Ribes aureum Pursh, Fl. i, 164 (1814).

In the foothills: Hot Springs, altitude 1,100 m., June 11 and August 3; Fall River Falls, altitude 1,000 m., June 17 (No. 690).

CRASSULACEÆ.

Sedum stenopetalum Pursh, Fl. i, 324 (1814).

It is nearly always more or less branched from the root. On rocky hills: Piedmont, altitude 1,100 m., June 27; Lead City, altitude 1,700 m., July 6; Buckhorn Mountain, near Custer, altitude 1,800 m., July 18 (No. 691).

HALORAGIDACEÆ.

Callitriche palustris L. Sp. Pl. ii, 969 (1753).

I mistook this for *Elatine americana*, which the plant very much resembles. Common in springs and brooks; spring, near Buckhorn Mountain, altitude 1,700 m., July 15; brook, 6 miles northwest of Custer, July 25 (No. 571).

ONAGRACEÆ.

Epilobium angustifolium L. Sp. Pl. i, 347 (1753).

Woody hills: Rochford, altitude 1,700 m., July 11; Sips Spring, in the Limestone District, altitude 1,800 m., July 28 (No. 692).

Epilobium lineare Muhl. Cat. 39 (1813).

The leaves are often opposite and are acutish, short-petioled, and without veins. In a marsh near Pringle, altitude 1,500 m., August 6 (No. 693).

Epilobium palustre L. Sp. Pl. i, 348 (1750).

This was named thus doubtfully by Dr. William Trelease. In his letter respecting the specimens submitted he adds: " However, they are pretty clearly that species, or possibly a hybrid of *lineare*, with the leaves broadened by hybridity." The latter seems to be the case. They are, very likely, hybrids of the preceding and *E. adenocaulon*, together with which two they were growing. After a careful search in the marsh, I could not find more than four specimens, nor did I see it elsewhere in the Hills. Pringle, altitude 1,500 m., August 6 (No. 694).

Epilobium adenocaulon Haussk. Oest. Bot. Zeitsch. 119 (1879).

Two forms were collected. One is branched, with smaller, more dentate leaves, approaching *E. coloratum* in general habit. Pringle, altitude 1,500 m., August 6 (No. 695).

The other is simple, with larger oblong-ovate leaves Rochford, altitude 1,600 m., July 11; Custer, altitude 1,600 m., July 14; Hot Springs, altitude 1,050 m., August 8 (No. 696).

Epilobium drummondii Haussk. Monogr. Gatt. Epil. 271 (1884).

Two forms were met with. One was tall, sometimes 5 dm. high, stringy, with narrow leaves: Rochford, altitude 1,700 m., July 12 (No. 697). The other was lower, with broader, ovate-lanceolate leaves sinuately toothed, sessile and half-clasping. Sips Spring, in the Limestone District, altitude 1,800 m., July 28 (No. 698).

Epilobium hornemanni Reichenb. Icon. Crit. ii, 73 (1824).

Only a few depauperate plants collected at Sips Spring together with the preceding (No. 699).

Epilobium paniculatum Nutt.; Torr. & Gr. Fl. i, 490 (1840).

Custer, altitude 1,700 m., August 21 (No. 571).

Gayophytum ramosissimum Torr. & Gr. Fl. i, 513 (1840).

Hills: Rochford, altitude 1,700 m., July 12; Bull Springs, altitude 1,900 m., July 26 (No. 700).

Œnothera biennis L. Sp. Pl. i, 346 (1753).

This is evidently native in western Nebraska, as well as in the Black Hills. In general habit it differs much from *O. biennis* of Europe. Rochford, altitude 1.600 m., July 11 (No. 701).

Another form, somewhat like the preceding, but not strigose, was also found. The pubescence is fine, silky, appressed; radical leaves many, obovate, about 2.5 cm. long; calyx tube nearly 4 cm. long, lobes linear-lanceolate about 1.25 cm. long; petals broadly obovate; pod linear-oblong, only a little narrower upward. Custer, altitude 1,700 m., July 15 (No. 702).

Œnothera sinuata L. Mant. 228 (1767).

Only a few small specimens secured at Hot Springs, altitude 1,100 m., June 14 (No. 703).

Œnothera albicaulis Pursh, Fl. ii, 733 (1814), not Nutt.; *Œ. pinnatifida* Nutt. Gen. i, 245 (1818).

The plant is very variable: Hot Springs, altitude 1,100 m., June 16: Hermosa, altitude 1,000 m., June 23: Custer, altitude 1,700 m., July 16 (No. 704).

Œnothera pallida leptophylla (Nutt.) Torr. & Gr. Fl. i, 495 (1838); *Œ. leptophylla* Nutt.; Torr. & Gr. loc. cit., as synonym; *Œ. albicaulis* Nutt. Gen. i, 245 (1818), not Pursh.

The typical *Œ. pallida* has more or less runcinate-toothed leaves. Broken soil: Custer, altitude 1,700 m., July 15 (No. 705).

Œnothera coronopifolia Torr. & Gr. Fl. i, 495 (1840).

Sandy soil: Custer, altitude 1,700 m., June 4: Hot Springs, altitude 1,100 m., June 15 (No. 706).

Œnothera cæspitosa Nutt. Fraser's Cat. (1813).

Battle Mountain, east of Hot Springs, altitude 1,200 m., June 15 (No. 707).

Œnothera serrulata Nutt. Gen. i, 246 (1818). ¹

Fall River Falls, altitude 1,000 m., June 17; Rapid City, altitude 1,050 m., June 25; Elk Canyon, altitude 1,200 m., June 29 (No. 708).

Gaura coccinea Pursh, Fl. ii, 733 (1814).

Hot Springs, altitude 1,100 m., June 13 (No. 709).

A form, perfectly smooth, with white bark which peels off like that of *Œnothera pallida*, was collected at Custer, altitude 1,700 m., July 15 (No. 710). This form is also found in western Nebraska, where I collected it in 1890 and 1891.

Gaura parviflora Dougl.; Hook. Fl. Bor. Amer. i, 208 (1834).

Hot Springs, altitude 1,050 m., August 2 (No. 711).

---

¹ Gray includes in his list, in Newton & Jenney's Report, also *Œ. chrysantha* Mx. (*Œ. pumila*), which must be an error, as that plant is strictly an Atlantic coast species.

Circæa alpina L. Sp. Pl. i, 9 (1753).

The specimens have less toothed leaves than usually, suggesting *C. pacifica;* but evidently they are not distinct from *C. alpina.* Shaded, damp place at the foot of Buckhorn Mountain, altitude 1,700 m., July 16 (No. 712).

LOASACEÆ.

Mentzelia decapetala (Pursh) Urban & Gilg, in Engler and Prantl. Nat. Pfl. iii Teil, 6 Abt. a. 111 (1891); *Bartonia decapetala* Pursh; Sims, Bot. Mag. xxxvi, t. 1487 (1812).

Bartonia ornata [1] is two years later and must give way to the older name. Rare: Hot Springs, altitude 1,100 m., August 9 (No. 713).

Mentzelia nuda (Pursh) Torr. & Gr. Fl. i, 535 (1840); *Bartonia nuda* Pursh, Fl. i, 328 (1814).

Rare: Hot Springs, altitude 1,050 m., August 3 (No. 714).

Mentzelia oligosperma Nutt.; Sims, Bot. Mag. xlii. t. 1760 (1815).

This was collected by Albert F. Woods near Hot Springs, but was not obtained by the writer.

CACTACEÆ.

Cactus missouriensis (Sweet) Kuntze, Rev. Gen. Pl. i, 259 (1891); *Mamillaria missouriensis* Sweet, Hort. Brit. 171 (1827).

Custer, altitude 1,650 m., June 4; Hot Springs, altitude 1,100 m., June 15 (No. 715).

Opuntia humifusa Raf. Ann. Nat. 15 (1820); *O. rafinesquii* Engelm. Pac. R. Rep. iv, 41 (1854).

Hills, 15 miles east of Custer, altitude 1,400 m., July 23 (No. 716).

Opuntia fragilis (Nutt.) Haw. Syn. Pl. Succ. Suppl. 82 (1819); *Cactus fragilis* Nutt. Gen. i, 296 (1818).

Plant only, collected near Minnekahta, altitude 1,300 m., August 5 (No. 717).

UMBELLIFERÆ.

Adorium tenuifolium (Nutt.) Kuntze, Rev. Gen. Pl. i. 264 (1891); *Musenium tenuifolium* Nutt.; Torr. & Gr. Fl. i, 612 (1840).

Very variable in size, according to the locality. The specimens from Custer, altitude 1,700 m., June 4 and August 1, seem to be typical, therefore like those of western Nebraska. Those from Hot Springs, altitude 1.100 m., June 13, growing in a more shaded locality, among gypsum rocks, are large with more striate scape. Those from the exposed granite rocks near Harneys Peak, altitude 2,200 m., June 8, are tufted, very low and delicate (No. 718).

Adorium hookeri (Torr. & Gr.); *Musenium divaricatum hookeri* Torr. & Gr. Fl. i, 612 (1840); *Musenium trachyspermum* Nutt.; Torr. & Gr., loc. cit., lower on the page. Hermosa, altitude 1,000 m., June 22 (No. 719).

Carum carui L. Sp. Pl. i, 263 (1753).

The ribs of the fruit have each a bundle of strengthening cells and a small oil tube, a fact that I have not seen pointed out.

Near a small brook, 3 miles north of Deadwood, altitude 1,400 m., July 5 (No. 720).

Carum gairdneri (Nutt.) Benth. & Hook. Gen. Pl. i, 891 (1867); *Edosmia gairdneri* Nutt.; Torr. & Gr. Fl. i, 612 (1840).

This is included in Gray's list, in Newton & Jenney's Report,[2] but no specimens have been seen by the author from the region.

---

[1] Pursh, Fl. i, 326 (1814).

[2] Geol. Surv. Black Hills, 533 (1880).

Zizia cordata (Walt.) Koch. Gen. Trib. Pl. Umb. 129 (1825); *Smyrnium cordatum* Walt. Fl. Car. 114 (1788).

This is the *Thaspium trifoliatum* of Newton & Jenney's Report.

Custer, altitude 1,650 m., June 2; Little Elk Canyon, altitude 1,100 m., June 28; Elk Canyon, altitude 1,200 m., June 29 (No. 721).

Berula erecta (Huds.) Coville, Contr. Nat. Herb. iv, 115 (1893); *Sium erectum* Huds. Fl. Angl. 103 (1762).

Hot Springs, altitude 1,050 m., August 10 (No. 722).

Cicuta virosa maculata (L.) Coult. & Rose, Rev. Umb. 130 (1888); *Cicuta maculata* L. Sp. Pl. i, 256 (1753).

Little Elk Canyon, altitude 1,100 m., June 28 (No. 723).

Osmorrhiza nuda Torr. Pac. R. Rep. iv, 93 (1856).

Elk Canyon, altitude 1,500 m., June 29; Lead City, altitude 1,600 m., July 6 (No. 724).

Osmorrhiza aristata (Thunb.) Rydberg, Bot. Surv. Neb. iii, 37 (1894); *Chaerophyllum aristatum* Thunb. Fl. Jap. 119 (1784); *O. longistylis* DC. Prodr. iv, 232 (1830).

Elk Canyon, altitude 1,200 m., June 29; Lead City, altitude 1,600 m., July 6; Hot Springs, altitude 1,100 m., (No. 725).

Cymopterus montanus Torr. & Gr. Fl. i, 624 (1840).

In fruit only: Hot Springs, altitude 1,100 m., June 14 (No. 726).

Cymopterus acaulis (Pursh) Rydberg, Bot. Surv. Neb. iii, 38 (1894); *Selinum acaule* Pursh, Fl. ii, 732 (1814); *Thapsia glomerata* Nutt. Gen. i, 184 (1818).

Only one specimen, collected while the train stopped at Edgemont, altitude 1,053 m., May 27 (No. 727).

Peucedanum villosum Nutt.; Wats. Bot. King Surv, v, 131 (1871).

Very common around Hot Springs, altitude 1,100 m., June 13 (No. 728)

Pastinaca sativa L. Sp. Pl. i, 262 (1753).

Escaped, along Rapid Creek, 6 miles above Rapid City, altitude 1,100 m., July 25 (No. 729).

Heracleum lanatum Mx. Fl. Bor. Amer. i, 166 (1803).

Rapid Creek above Rapid City, altitude 1,100 m., June 25; Little Elk, altitude 1,100 m., June 28 (No. 730).

Sanicula canadensis L. Sp. Pl. i, 235 (1753).

Elk Canyon, altitude 1,200 m., June 29; Lead City, altitude 1,600 m., July 6 (No. 731.)

ARALIACEÆ.

Aralia nudicaulis L. Sp. Pl. i, 274 (1753).

Little Elk Canyon, altitude 1,200 m., June 28; Elk Canyon, altitude 1,300 m., June 29; Lead City, altitude 1,700 m., July 6; Custer, altitude 1,600 m., August 15 (No. 732).

At one place in Little Elk Canyon all specimens differed from the usual form in being lower, the leaves being only 2 dm. or less long, while in the ordinary form they are 3 dm.; the umbels 1 to 4 on erect branches (in the ordinary form they are more or less spreading): flowers larger and blooming before the leaves are fully developed. (No. 733).

CORNACEÆ.

Cornus canadensis L. Sp. Pl. i, 118 (1753).

Lead City, altitude 1,600 m., July 4; Sylvan Lake, altitude 2,000 m., August 17 (No. 734).

Cornus baileyi Coult. & Evans. Bot. Gaz. xv, 37 (1890).

I refer this here with doubt, as I did not see it in fruit. The pubescence of the peduncles is woolly, and that of the lower surface of the leaves is more or less mixed with woolly hairs. The leaves are narrower than in the next, of which it may be a more hairy variety, and the bark of the branches is browner and pubescent. It may also be a narrow-leafed form of *C. pubescens*.

Little Elk Canyon, altitude 1,200 m., June 28; Rochford, altitude 1,700 m., July 11 (No. 735).

Cornus stolonifera Mx. Fl. i, 92 (1803).

The bark is purplish red, the pubescence silky. In fruit only: Sylvan Lake, altitude 1,900 m., August 17 (No. 736).

CAPRIFOLIACEÆ.

Adoxa moschetellina L. Sp. Pl. i, 367 (1753).

Only a few small specimens in bud were secured, below Sylvan Lake, altitude 2,000 m., June 3, and a few in fruit, badly damaged by rust, under a rock near Sips Spring, in the Limestone District, altitude 1,900 m., July 28 (No. 737).

Sambucus racemosa L. Sp. Pl. i, 270 (1753).

Three forms belonging here were collected. In one the annual shoot and the peduncles are more or less roughish-pubescent and warty; leaves oblong-lanceolate, long acuminate, closely serrate; cyme many-flowered, roundish. The mature fruit was not seen. The cyme resembles that of *S. melanocarpa*, but the leaves are more like those of *S. racemosa* and the flowers are "dull white, drying brownish." Canyon, north of Runkels, altitude 1,300 m., July 30; Custer, altitude 1,700 m., July 14 (No. 738).

A form which I think is more typical has the shoots perfectly smooth and light colored, leaflets ovate-lanceolate with shorter acumination, cyme longer, but smaller; fruit bright red as in *Shepherdia argentea*. Sylvan Lake, altitude 2,000 m., August 17 (No. 739).

Together with this form was another in every respect like it, except that the berries were amber yellow, resembling and being the analogue of the amber-colored variety of Shepherdia growing in western Nebraska (No. 740).

Viburnum opulus L. Sp. Pl. i, 268 (1753).

Canyon north of Runkels, altitude 1,300 m., June 30 (No. 741).

Viburnum lentago L. Sp. Pl. i, 268 (1753).

The margined petioles are often rufous-pubescent as they should be in *V. pruni-folium*, which is said to grow in the Black Hills. I think, however, that this has been mistaken for that. Little Elk, altitude 1,200 m., June 28; Runkels, altitude 1,300 m., June 30 (No. 742).

Linnæa borealis L. Sp. Pl. ii, 631 (1753).

Common in the Northern Hills, but also seen in the Harney Range. Elk Canyon, altitude 1,200 m., June 29; Lead City, altitude 1,600 m., July 6 (No. 743).

Symphoricarpos racemosus pauciflorus Robbins, Gray, Man. ed. 5, 203 (1867).

Common: Little Elk Canyon, altitude 1,200 m., June 28; Elk Canyon, altitude 1,300 m., June 29; Lead City, altitude 1,600 m., July 6; Rochford, altitude 1,700 m., July 12 (No. 744).

Symphoricarpos occidentalis Hook. Fl. Bor. Amer. i, 285 (1834).

Very variable; flowers few or many; style glabrous or sparsely villose; stout or slender; leaves entire or lobed, large and thick or small and thin. Some forms approaching the preceding. Elk Canyon, altitude 1,200 m., June 29; Hot Springs, altitude 1,100 m., August 9; Custer, altitude 1,700 m., August 12 (No. 745).

A form with thin, ovate, acute or acuminate, or even pointed leaves, and a thin, less bearded corolla was collected at the last place, August 19 (No. 746). The same has been collected by Dr. Chas. E. Bessey in Colorado.

Lonicera hirsuta glaucescens, var. nov.; *Lonicera parviflora* var.? Torr. & Gr. Fl. ii, 7 (1840), partly; *Lonicera douglasii* Hook. Fl. Bor. Amer. i, 282 (1833) (?), not *Caprifolium douglasii* Lindl. Trans. Hort. Soc. vii, 244, which is *L. hirsuta* proper.

Leaves 3 to 5 cm. long and 1.5 to 3 cm. wide, smooth above, slightly glaucous beneath, not ciliate, generally only the upper pair connate; corolla about 1.5 cm. long, pubescent on the outside, strongly gibbose at the base of the tube, yellow, changing into reddish; stem smooth; bark first green, afterwards grayish straw-colored, more or less shreddy.

It differs from the true *L. hirsuta* in the smaller leaves, which are perfectly smooth above and decidedly glaucous beneath (in *L. hirsuta* they are seldom glaucous), in the corolla, which is more gibbose, and in the smooth and shreddy stem. It may be a distinct species. It resembles somewhat *L. glauca*, from which it differs in the longer, hairy, and gibbose corolla. It is sometimes a low shrub, sometimes high-climbing. The only specimen in the National Herbarium, except those from the Black Hills, was collected by S. M. Tracy in the Rocky Mountains in 1888 (no locality given). In the herbarium of Harvard University are the following specimens: In Dr. Gray's collections, a fragmentary one labeled "*Lonicera douglasii* Fl. Bor. Am. Hooker misit January, 1835;" and another labeled "*L. hirsuta* var. *douglasii* Hooker approaching *glauca*." The others were probably received later, as there are no indications that Dr. Gray had ever studied them, viz: *Lonicera glauca* Hill, Agricultural College, Ingham County, Mich., 1860 (no collector given); Ex. Herb. Thurber, coll. T. J. Hale, *Lonicera douglasii* DC., hab. Ripon, Wis., 1861; 150 *Lonicera hirsuta* (*L. douglasii* Hook). "River That Turns," July 13, 1879, coll. J. Macoun, F. L. S. Railway Survey; 96½ Herb. of Wm. Werner, *Lonicera glauca* Hill [corrected to], *L. hirsuta* Eaton, Painesville, Ohio, 1890; Herb. L. H. Bailey, jr., *Lonicera glauca* Hill, Lansing, Mich., 1886.

Little Elk, altitude 1,200 m., June 28; Lead City, altitude 1,600 m., July 6 (No. 747).

RUBIACEÆ.

Galium aparine L. Sp. Pl. i, 108 (1753).
Hot Springs, altitude 1,050 m., June 13 (No. 748).

Galium triflorum Mx. Fl. i, 80 (1803).
Elk Canyon, altitude 1,200 m., June 29; Lead City, altitude 1,600 m., July 6 (No. 749).

Galium boreale L. Sp. Pl. i, 108 (1753).
Hot Springs, altitude 1,100 m., June 15; Hermosa, altitude 1,100 m., June 22; Rapid City, altitude 1,000 m., June 25; Elk Canyon, altitude 1,200 m., June 29 (No. 750).

VALERIANACEÆ.

Valeriana edulis Nutt.; Torr. & Gr. Fl. ii, 48 (1841).
Rochford, altitude 1,700 m., July 12; Oreville, altitude 1,700 m., July 16 (No. 751).
Valeriana sylvatica Banks; Richards. Bot. App. 730 (1823).
Rochford, altitude 1,600 m., June 12; Oreville, altitude 1,700 m., July 16 (No. 752).

COMPOSITÆ.

Lacinaria scariosa (L.) Hill, Veg. Syst. iv, 49 (1762); *Serratula scariosa* L. Sp. Pl. ii, 818 (1753).
Custer, altitude 1,600 m., Aug. 1 (No. 753).

Lacinaria punctata (Hook.) Kuntze, Rev. Gen. Pl. i, 349 (1891); *Liatris punctata* Hook. Fl. Bor. Amer. i, 306 (1833).
Custer, altitude 1,600 m., August 1; Hot Springs, altitude 1,100 m., August 3 (No. 754).

Kuhnia glutinosa Ell. Bot. S. Car. & Georg. ii, 292 (1821-1824); *K. eupatorioides glutinosa* Hitchcock, Trans. St. Louis Acad. v, 498 (1891).
Hot Springs, altitude 1,100 m., August 2, (No. 755).

Grindelia squarrosa (Pursh) Dunal, in DC. Prodr. v, 315 (1836); *Donia squarrosa* Pursh, Fl. ii, 559 (1814).

Hot Springs, altitude 1,100 m., August 2 (No. 756).

Two depauperate specimens in bloom (several plants were seen) were collected above Fall River Falls, altitude 1,000 m., June 17. These two had narrow leaves, more toothed than usual, and I took them at first to be *G. nana* Nutt., but they must belong to *G. squarrosa* (No. 757).

Chrysopsis villosa (Pursh) Nutt. Gen. ii, 151 (1818); *Amellus villosus* Pursh, Fl. ii, 564 (1814).

Sandy soil: Rochford, altitude 1,600 m., July 12; Custer, altitude 1,650 m., July 12, (No. 758).

Chrysopsis villosa canescens (DC.) Gray, Syn. Fl. i, pt. ii, 123 (1884); *Aplopappus* (?) *canescens* DC. Prodr. v, 349 (1836).

Custer, altitude 1,650 m., July 15; Hot Springs, altitude 1,100 m., August 2 (No. 759).

Eriocarpum grindelioides Nutt. Trans. Amer. Phil. Soc. ser. 2, vii, 321 (1841); *Aplopappus nuttallii* Torr. & Gr. Fl. ii, 242 (1842). Gypsum rocks above Hot Springs, altitude 1,100 m., August 2 (No. 760).

Eriocarpum spinulosum (Pursh) Greene, Erythea, ii, 108 (1894); *Amellus spinulosus* Pursh, Fl. ii, 564 (1814): *Aplopappus* (?) *spinulosus* DC. Prodr. v, 347 (1836).

Rare: Hot Springs, altitude 1,100 m., August 2 (No. 761).

Solidago erecta Pursh, Fl. ii, 512 (1814).

This is the *S. speciosa angustata* Torr. & Gr., of Newton & Jenney's Report. I refer it doubtfully here. It does not agree with the original description by Pursh, being perfectly smooth except the margins of the leaves, which are scabrous. It agrees well with the description of *S. erecta* by Elliott.[1] Gray[2] regards the two as the same species notwithstanding the pubescence attributed to the first. It is near *S. speciosa*, but the primary veins are more or less prominent and often looped.

Hilly places: Custer, altitude 1,700 m., August 16 (No. 762).

Solidago missouriensis Nutt. Journ. Acad. Phil. vii, 32 (1834).

Very variable. The form held as the typical one, that is, stouter with spreading panicle of recurved branches, was collected at Custer, altitude 1,700 m., August 14 (No. 763). This is, however, not the original *S. missouriensis*, but should, if held separate, be called variety *glaberrima* (*S. glaberrima* Martens). The true *S. missouriensis* was also collected, viz. at Custer, altitude 1,700 m., August 16 (No. 764). This was named variety *montana* by Dr. Gray. Another form was found with broad leaves, the lower often 1.5 cm. wide, the stem tall, 5 dm. high, panicle with upright branches and large heads. It may be the variety *extraria* Gray, or, perhaps, a hybrid of *S. missouriensis* and the preceding. together with which it grew. Custer, August 16 (No. 765).

Solidago rupestris Raf. Ann. Nat. 14 (1820).

Owing to the meager material, the determination is doubtful. It may perhaps be a narrow and thin-leafed form of *S. serotina* Ait. Little Elk, altitude 1,100 m., June 28 (No. 766).

Solidago canadensis L. Sp. Pl. ii, 878 (1753).

Custer, altitude 1,650 m., August 1 (No. 767).

Solidago canadensis procera (Ait.) Torr. & Gr. Fl. ii, 224 (1842); *S. procera* Ait. Hort. Kew. iii, 211 (1789).

Custer, altitude 1,700 m., August 1 (No. 1207).

Solidago nemoralis Ait. Hort. Kew. iii, 213 (1789).

My specimens are more or less scabrous, some have axillary clusters and resemble somewhat *S. bicolor concolor*. Custer, altitude 1,700 m., August 16 (No. 768).

---

[1] Bot. S. Car. & Georg. i, 385 (1817).
[2] Proc. Amer. Acad. viii, 308 (1870).

Solidago rigida L. Sp. Pl. ii, 880 (1753).

Much smaller than the form common in eastern Nebraska, 3 to 6 dm. high, more cinereous; corymb more open and heads smaller; radical and lower cauline leaves gradually acuminating into a winged petiole, all with a clasping but not decurrent base. Custer, dry land, altitude 1,700 m., August 16 (No. 769).

Euthamia graminifolia (L.) Nutt. Gen. ii, 162 (1818); *Chrysocoma graminifolia* L. Sp. Pl. ii, 841 (1753); *Solidago lanceolata* L. Mant. 114 (1767).

The corymbs in all specimens seen in the Black Hills are unusually small and dense with larger heads. Hot Springs, altitude 1,100 m., August 10 (No. 770).

Aster sibiricus L. Sp. Pl. ii, 872 (1753).

Shaded hillsides: Custer, altitude 1,700 m., August 13; Rochford, altitude 1,700 m., July 12 (No. 771). All specimens are unusually low.

Aster lævis L. Sp. Pl. ii, 876 (1753).

All the specimens are small, some very low and slender, with narrowly lanceolate leaves. Custer, altitude 1,700 m., August 16 (No. 772).

Aster multiflorus incanopilosus (Lindl.) Rydb. Contr. Nat. Herb. iii, 163 (1895); *A. ramulosus incanopilosus* Lindl.; Hook. Fl. Bor. Amer. ii, 13 (1831); *A. multiflorus commutatus* Torr. & Gr. Fl. ii, 121 (1841).

Custer, altitude 1,600 m., August 16 (No. 773).

Aster patulus Lam. Encycl. i, 308 (1783).

Like *A. prenanthoides* Muhl., but the base of the leaves is not cordate at all. Only a few specimens secured. Custer, altitude 1,700 m., August 13 (No. 774).

Aster junceus Ait. Hort. Kew. iii, 204 (1789).

See my remark on this species, this volume, p. 163. The leaves are linear, 1-nerved, with slightly revolute margins.

Wet meadow : Custer, altitude 1,600 m., August 15 (No. 775).

Aster salicifolius Lam. Encyl. i, 306 (1783).

Meadow : Custer, altitude 1,600 m., August 15 (No. 776).

Aster ptarmicoides (Nees) Torr. & Gr. Fl. ii, 160 (1841); *Doellingeria ptarmicoides* Nees. Gen. & Sp. Ast. 183 (1832).

Limestone District, near Bull Springs, altitude 1,900 m., July 25 (No. 777).

Aster falcatus Lindl.; DC. Prodr. v, 241 (1836).

In Gray's list, Newton & Jenney's Report[1]; no specimens from this region seen by the writer.

Aster paniculatus Lam. Encycl. i, 306 (1783); *A simplex* Willd. Enum. 887 (1809). In Gray's list only.[2]

Aster tanacetifolius H. B. K. Nov. Gen. iv, 95 (1820).

In Gray's list only.[2]

Erigeron asper Nutt. Gen. ii, 147 (1818).

More or less strigose all over. Generally 3 to 5 dm. high, with several heads; but in higher altitudes they are only 2 to 2.5 dm. high, with 1 to 3 heads. Former state, Rochford, altitude 1,600 m., July 11; latter state, Limestone District, July 26 (No. 778).

Erigeron subtrinervis Rydberg, Mem. Torr. Club, v. 328 (1894); *E. glabellus mollis* Gray, Proc. Acad. Phila. 1863, 61 (1863), not *E. mollis* D. Don, Prodr. Fl. Nep. 172 (1825).

Plant hairy throughout, from grayish scabro-strigose to soft-pubescent, leafy to the top; leaves thin, more or less distinctly triple-nerved, the lower oblanceolate, petioled, the upper oblong to ovate-lanceolate, sessile and half-clasping; heads 1 to 5, corymbose, 1.5 cm. wide and 0.75 cm. high; rays 0.75 to 1 cm. long, blue or flesh color; involucre hirsute.

---

[1] Geol. Surv. Black Hills, 531 (1880).   [2] Loc. cit.

Gray (loc. cit.) says: "From the shape of the leaves, and their size and abundance up to the summit of the stem, this should rather be referred to *E. macranthum;* but the pubescence is strange for that species;" and in the Synoptical Flora he adds, "Perhaps a distinct species." I believe it to be distinct and rather more related to *E. macranthus* than to *E. glabellus:* in fact, it differs little from the former except in the pubescence, which is not only more copious but of a different nature, more resembling that of the hirsute forms of *E. glabellus.* In *E. macranthus* the bracts are nearly smooth, somewhat minutely glandular or puberulent: in *E. subtrinervis* they are covered with longer spreading hairs. I have named the species *E. subtrinervis,* from the fact that the lower lateral veins are often stronger, making the leaves look as if triple-nerved, a character often seen in *E. speciosus,* sometimes in *E. macranthus,* though I have not seen it in *E. glabellus.* The leaves are generally thinner than in any of these three related species.

In woods: Custer, altitude, 1,700 m., August 16 (No. 779).

Erigeron pumilus Nutt. Gen. ii, 147 (1818).
Dry table-lands: Hot Springs, altitude 1,100 m., June 13 (No. 780).

Erigeron compositus Pursh, Fl. ii, 535 (1814).
All my specimens are strictly scapose and densely matted. Exposed rocks in the Limestone District, altitude 1,900 m., July 27. Also seen near the Needles, altitude about 2,100 m., in the Harney Range (No. 781).

Erigeron canus Gray, Pl. Fendl. 67 (1849).
Dry table-lands: Hot Springs, altitude 1,100 m., June 13 (No. 782).

Erigeron philadelphicus L. Sp. Pl. ii. 863 (1753).
Near water: Hot Springs, altitude 1,050 m., June 17: Hermosa, altitude 1,000 m., June 22; Elk Canyon, altitude 1,200 m., June 29 (No. 783).

Erigeron flagellaris Gray, Pl. Fendl. 68 (1849).
Rich soil: Hot Springs, altitude 1,050 m., June 17; Hermosa, altitude 1,000 m., June 22; Buckhorn Mountain, near Custer, altitude 1,800 m., July 14. The specimens from the latter place, have much smaller radical leaves than the rest (No. 784).

Erigeron ramosus beyrichii (Fisch. & Mey.) Smith & Pound, Bot. Surv. Nebr. ii, 11 (1893); *Stenactis beyrichii* Fisch. & Mey. Ind. Sem. Petrop. 27 (1824).
Below Buckhorn Mountain, Custer, altitude 1,700 m., August 14 (No. 785).

Erigeron armerifolius Turcz.; DC. Prodr. v, 291 (1836).
Radical leaves numerous, spatulate. Wet meadow, Custer, altitude 1,700 m., August 20 (No. 786).

Erigeron canadensis L. Sp. Pl. ii. 863 (1753).
This is rare in the Black Hills. Custer, altitude 1,650 m., August 20 (No. 787). A depauperate form, resembling much *E. divaricatus,* was collected in Ruby Glen, Custer, altitude 1,700 m., August 20 (No. 788).

Filago prolifera (Nutt.) Britton, Mem. Torr. Club, v, 329 (1891); *Evax prolifera* Nutt.: DC. Prodr. v, 459 (1836).
Barren Hills: Hot Springs, altitude 1,100 m., June 13 (No. 789).

Antennaria dioica (L.) Gærtn. Fruct. ii, 410 (1791); *Gnaphalium dioicum* L. Sp. Pl. ii, 850 (1753).
The typical *A. dioica* has leaves 1 to 2 cm. long, and 0.3 to 0.6 cm. wide, silvery white on both sides; bracts all obtuse, the papery portion white or pinkish red. Borders of woods: Hot Springs, altitude 1,100 m., June 13; Rochfort, altitude 1,600 m., July 11 (No. 790).
A peculiar form with several crowded heads, and often acute bracts, I also refer here. The bracts are of livid-brownish color, approaching *A. alpina* in this respect, but the plant is in every respect larger. Custer, altitude 1,700 m., July 4 (No. 792).

Antennaria dioica parvifolia (Nutt.) Torr. & Gr. Fl. ii, 431 (1840); *Antennaria parvifolia* Nutt. Trans. Amer. Phil. Soc. n. ser. vii, 406 (1841).

Radical leaves narrow, oblanceolate, more or less revolute, finely silvery on both sides; flowering stem slender, bearing smaller heads with brightly rose-colored bracts. The plant is more diffusely spreading. Prairie: Rochford, altitude 1,600 m., July 12 (No. 791).

Antennaria plantaginifolia (L.) Richards. Bot. App. ed. 2, 30 (1823); *Gnaphalium plantaginifolium* L. Sp. Pl. ii, 850 (1753).

This species is very variable, at least if all the western forms belong to it. The typical form with large, thin, 3-ribbed leaves was not met with. What I take for an alpine form of this species was collected early in the spring, around Custer, altitude 1,650 m., May 30 (No. 791). This is low, 7 to 10 cm. high, with smaller heads, radical leaves obovate or oblong, about 3 cm. long, glabrous above, white beneath. The other two forms collected belong to the western form of *A. plantaginifolia*, which perhaps is distinct from the eastern. The leaves are smaller, 1.5 to 1 cm. long, silky on both sides, and seldom 3-nerved. In one of the forms the leaves are about 3 cm. long, the stem more robust and more floccose. Hot Springs, altitude 1,100 m., June 13; Hermosa, altitude 1,050 m., June 2; Lead City, altitude 1,700 m., July 1 (No. 793). The other form is more slender and less floccose, the leaves about one-half the size of those of the last. This is the same as No. 173 of my western Nebraska collection. Lead City, altitude 1,700 m., July 6 (No. 795).

Antennaria margaritacea (L.) Hook. Fl. Bor. Amer. i, 329 (1833); *Gnaphalium margaritaceum* L. Sp. Pl. i, 85 (1753).

Englewood, altitude 1,600 m., August 13 (No. 796).

Iva xanthifolia Nutt. Gen. ii. 185 (1818).

Waste places, rare: Custer, altitude 1,600 m., August 19 (No. 797).

Ambrosia psilostachya DC. Prodr. v, 526 (1836).

Rare: Hot Springs, altitude 1,700 m., August 2 (No. 798).

Ambrosia artemisiæfolia L. Sp. Pl. ii. 988 (1753).

Custer, altitude 1,650 m., August 11 (No. 799).

Brauneria pallida (Nutt.) Britton, Mem. Torr. Club, v, 333 (1895); *Rudbeckia pallida* Nutt. Journ. Acad. Phila. v, 77 (1834); *Echinacea angustifolia* DC. Prodr. v, 554 (1836).

Prairie: Custer, altitude 1,600 m., August 1 (No. 800).

Rudbeckia hirta L. Sp. Pl. ii, 907 (1753).

Meadows: Rochford, altitude 1,600 m., July 12; in the Limestone District, altitude 1,900 m., July 26 (No. 801). One specimen was very leafy, with narrow linear-lanceolate leaves.

Lepachys columnaris (Pursh) Torr. & Gr. Fl. ii, 315 (1841); *Rudbeckia columnaris* Pursh, Fl. ii, 575 (1814).

Rare: Only a few specimens secured at Custer, altitude 1,650 m., August 1 (No. 802).

Helianthus annuus L. Sp. Pl. ii. 904 (1753).

Custer, altitude 1,600 m., August 1 (No. 803).

Helianthus petiolaris Nutt. Journ. Acad. Phila. ii, 115 (1821).

Only a few specimens were secured; all had some of the leaves opposite. Hot Springs, altitude 1,100 m., August 2 (No. 804).

Helianthus scaberrimus Ell. Bot. S. Car. & Georg. ii, 423 (1821); *H. rigidus* Desf. Cat. Hort. Paris, ed. 3, 184 (1829).

Custer, altitude 1,700 m., August 1 (No. 805).

Helianthus maximiliani Schrad. Ind. Sem. Hort. Goett. (1835).

Custer, altitude 1,700 m., August 15 (No. 806).

Helianthella quinquenervis (Hook.) Gray, Proc. Amer. Acad. xix, 10 (1883); *Helianthus quinquenervis* Hook. Lond. Journ. Bot. vi, 247 (1847).

The leaves in my specimens are as often alternate as opposite, and the plant is much taller than the specimens in the National Herbarium.

Hills: Limestone District near Sips Spring, altitude 1,900 m., July 27 (No. 807).

Balsamorhiza sagittata (Pursh) Nutt. Trans. Amer. Phil. Soc. vii, 350 (1841); *Buphthalmum sagittatum* Pursh, Fl. ii, 564 (1814).
It was collected in fruit only on hills near Sips Spring, altitude 1,900 m., July 27 (No. 808).

Bidens lævis (L.) B. S. P. Cat. Pl. N. Y. 29 (1888); *Helianthus lævis* L. Sp. Pl. ii, 906 (1753).
All the specimens collected were low, with short rays. Pringle, altitude 1,500 m., August 6 (No. 809). Some were very slender; leaves about 2 to 3 cm. long, not connate; heads about 6 mm. long, few-flowered; sparingly strigose (No. 810).

Hymenopappus filifolius Hook. Fl. Bor. Amer. i, 317 (1834).
Hot Springs, altitude 1,100 m., June 13 and August 3 (No. 811).

Ptilepida acaulis (Pursh) Britton. Mem. Torr. Club, v, 339 (1891); *Gaillardia acaulis* Pursh, Fl. ii, 743 (1814); *Actinella acaulis* Nutt. Gen. ii, 173 (1818).
Dry hills: Hot Springs, altitude 1,100 m., June 12; Lead City, altitude 1,800 m., July 4 (No. 812).

Gaillardia aristata Pursh, Fl. ii, 573 (1814).
Prairies: Rochford, altitude 1,600 m., July 12; Limestone District, altitude 1,900 m., July 26 (No. 813).

Dysodia papposa (Vent.) Hitch. Trans. St. Louis Acad. v. 503 (1891); *Tagetes papposa* Vent. Hort. Cels. t. 36 (1800).
Waste places: Hot Springs, altitude 1,050 m., August 8 (No. 814).

Anthemis cotula L. Sp. Pl. ii, 894 (1753).
Hot Springs, altitude 1,050 m., August 2 (No. 815).

Achillea millefolium L. Sp. Pl. ii, 899 (1753).
Elk Canyon, altitude 1,200 m., June 29; Lead City, altitude 1,600 m., July 4; Hot Springs, altitude 1,100 m., August 2 (No. 816).

Artemisia canadensis Mx. Fl. ii, 128 (1803).
Dry places: Custer, altitude 1.700 m., August 15 (No. 817).

Artemisia dracunculoides Pursh, Fl. ii, 712 (1814).
Custer, altitude 1,700 m., August 15 (No. 818).

Artemisia frigida Willd. Sp. Pl. iii, 1838 (1804).
Dry places: Custer, altitude 1,700 m., August 19 (No. 819).

Artemisia gnaphalodes Nutt. Gen. ii, 143 (1818).
Custer, altitude 1,700 m., August 19 (No. 820).

Petasites sagittata (Pursh) Gray, Bot. Cal. i, 407 (1876); *Tussilago sagittata* Pursh, Fl. ii, 531 (1814).
A large patch found on Rapid Creek, above Rochford, altitude 1,650 m., July 12, but only three specimens in fruit secured (No. 821).

Arnica cordifolia Hook. Fl. Bor. Amer. i, 331 (1834).
On shaded hillsides: Elk Canyon, altitude 1,200 m., July 29; Lead City, altitude 1,600 m., July 4; Rochford, altitude 1,700 m., July 12 (No. 822).

Arnica alpina (L.) Olin, Monogr. Arn. Ups. 1799; *Arnica montana alpina* L. Sp. Pl. ii, 884 (1753).
In the Black Hills this is generally 3-cephalous, and 3 to 5 dm. high. Little Elk Canyon, altitude 1,100 m., July 28; Custer, altitude, 1,800 m., July 16 (No. 823).

Senecio rapifolius Nutt. Trans. Amer. Phil. Soc. vii, 409 (1841).
The specimens are unusually large, some being 6 dm. high. In shaded places among rocks: Custer, altitude 1,700 m., August 12 (No. 824).

Senecio lugens Richards. App. Frankl. Journ. 747 (1823).
In the specimens collected the leaves are unusually narrow and nearly entire.

The plant may be a form of the next. Rochford, altitude 1,650 m., July 11; Hot Springs, altitude 1,050 m., August 8 (No. 825).

Senecio integerrimus Nutt. Gen. ii, 105 (1818).

This resembles a specimen collected by Mr. Nicollet, which according to Torrey and Gray [1] is *S. integerrimus* Nutt. The other specimens in the National Herbarium seem to belong to some other species. The type specimens of Nuttall I have not seen. The bracts and the base of the involucre are somewhat fleshy. Prairie, near Squaw Creek, Hermosa, altitude 1,050 m., June 23 (No. 826).

Senecio balsamitæ Muhl.; Willd. Sp. Pl. iii, 1998 (1801).

Leaves very thin and wholly glabrous, bright green, the lower obovate, serrate. Lead City, altitude 1,600 m., July 6; Rochford, altitude 1.600 m., July 11; Custer, altitude 1,700 m., July 16 (No. 827).

Senecio plattensis Nutt. Trans. Amer. Phil. Soc. ser. 2, vii, 413 (1841).

Hot Springs, altitude 1,100 m., June 18 (No. 828).

Senecio canus Hook. Fl. Bor. Amer. i, 333 (1831).

This is a very variable species. Three forms were collected, one of them the typical. This is 3 to 5 dm. high, densely woolly; root leaves entire, broadly oblanceolate, oblong, or spatulate; 7 to 10 cm. long, 1 to 1.5 cm. wide. Battle Mountain, near Hot Springs, altitude 1,200 m., June 18; Lead City, altitude 1,700 m., July 4 (No. 829). The second form is somewhat like the last but greener with deciduous wool and thin leaves. It approaches somewhat the preceding species, especially as to the involucre, which is nearly glabrous. Lead City, altitude 1,600 m., July 6 (No. 830). The third is a low form with narrow leaves, the lower narrowly spatulate or oblanceolate, 3 to 7 cm. long and about 4 mm. wide, white woolly, with more or less revolute margins. Dry table-land: Hot Springs, altitude 1,100 m., June 13 (No. 831).

Carduus undulatus Nutt. Gen. ii, 130 (1818); *Cnicus undulatus* Gray, Proc. Amer. Acad. x, 42 (1874).

Custer, altitude 1,650 m., August 12 (No. 832).

Carduus undulatus ochrocentrus, nom. nov.: *Cnicus undulatus ochrocentrus* Gray, Proc. Amer. Acad. x, 43 (1874); *Cirsium ochrocentrum* Gray, Pl. Fendl. 110 (1849); *Cnicus ochrocentrus* Gray, Proc. Amer. Acad. xix, 57 (1883).

This is the northern form referred to *Carduus ochrocentrus*. The southern, that is, the original form, differs in being stouter and more white with broader bracts. Whether the latter is distinct from *C. undulatus* I am not prepared to decide. The northern form is, I think, only a variety, and differs only in the prickles of the involucre, which are much longer and stouter. Intermediate forms are sometimes seen.

Custer, altitude 1,650 m., August 1 (No. 833).

Carduus drummondii (Gray); *Cnicus drummondii* Gray, Proc. Amer. Acad. x, 40 (1871).

Some of the specimens are very near to the variety *acaulescens* Gray. Meadows: Custer, altitude 1,650 m., July 16 (No. 834).

Centaurea cyanus L. Sp. Pl. ii, 911 (1753).

Only one specimen collected: Roadside, not far from Hot Springs, altitude 1,100 m., August 2 (No. 835).

Hieracium canadense Mx. Fl. ii, 86 (1803).

The specimens referred to this species have much thinner, narrower, and less dentate leaves than the common form. Custer, altitude 1,650 m., July 12 (No. 836).

Hieracium umbellatum L. Sp. Pl. ii, 804 (1753).

The specimens of this have fewer heads than usual. Wet meadow: Rapid City, altitude 1,000 m., June 25 (No. 837).

[1] Fl. ii, 439.

Hieracium sp.

A single specimen was found, together with the preceding, of a plant which resembles it as to the leaves, but has the stem hispid above, and the bracts broad (No. 838).

Hieracium fendleri Schultz Bip. Bonplandia, ix, 173 (1861).

The Black Hills are outside of the supposed range of this species, but the specimens agree fully with Gray's description and with specimens in the National Herbarium and the herbarium of the University of Nebraska. Rochford, on a dry hill, altitude 1,700 m., July 12 (No. 839).

Crepis runcinata (James) Torr. & Gr. Fl. ii, 487 (1843); *Hieracium runcinatum* James, Long Exped. i, 453 (1823).

The typical form, but with the leaves less runcinate. Hermosa, altitude 1,050 m., June 23; Whitewood, altitude 1,200 m., July 7; Rochford, altitude 1,700 m., July 11; Custer, altitude 1,700 m., July 16 (No. 840).

A stouter form, 4 to 5 dm. high, with broader leaves; heads larger; involucre of thicker bracts, with the peduncles and upper part of the stem densely covered with yellowish glands. A specimen of this form in the National Herbarium is labeled variety *hispidulosa* Howell, but I can not find any description. Meadow: Rapid City, altitude 1,000 m., July 25 (No. 841).

Prenanthes racemosa Mx. Fl. ii, 84 (1803).

Wet meadow: Custer, altitude 1,600 m., August 15 (No. 842).

Agoseris glauca (Pursh) Greene, Pittonia, ii, 176 (1891); *Troximon glaucum* Pursh, Fl. ii, 505 (1814).

Meadows: Rapid City, altitude 1,000 m., June 25; Elk Canyon, altitude 1,200 m., June 29 (No. 843).

Agoseris glauca parviflora (Nutt.); *Troximon parviflorum* Nutt. Trans. Amer. Phil. Soc. vii, 431 (1841).

Professor Greene regards this as a distinct species, stating that it differs in the leaves being strictly 2-ranked, a character which is, however, sometimes found in the true *A. glauca*.

Custer, altitude 1,650 m., July 16 (No. 844).

Agoseris scorsoneræfolia (Schrad.) Greene, Pittonia, ii, 177 (1891); *Ammogeton scorsoneræfolium* Schrad. Cat. Sem. Goett. 1 (1833).

Railroad embankment: Custer, altitude 1,650 m., July 16 (No. 845).

Taraxacum taraxacum (L.) Karst. Deutsch. Fl. 1138 (1880-1883); *Leontodon taraxacum* L. Sp. Pl. ii, 798 (1753).

Rapid City, altitude 975 m., July 25 (No. 846).

Lactuca ludoviciana (Nutt.) DC. Prodr. vii, 141 (1839); *Sonchus ludovicianus* Nutt. Gen. ii, 125 (1818).

Hillside east of Custer, altitude 1,600 m., July 23 (No. 847).

Lactuca pulchella (Pursh) DC. Prodr. vii, 134 (1839); *Sonchus pulchellus* Pursh, Fl. ii, 502 (1814).

Custer, altitude 1,650 m., August 16; Hot Springs, altitude 1,050 m., August 2 (No. 848).

Sonchus asper (L.) All. Fl. Ped. i, 222 (1785); *Sonchus oleraceus asper* L. Sp. Pl. ii, 794 (1753).

Hot Springs, altitude 1,050 m., August 2 (No. 849).

LOBELIACEÆ.

Lobelia spicata hirtella Gray, Syn. Fl. ii, pt. i, 6 (1878).

Prairie: Custer, altitude 1,600 m., August 1 (No. 850).

CAMPANULACEÆ.

Legouzia perfoliata (L.) Britton. Mem. Torr. Club, v, 309 (1894); *Campanula perfoliata* L. Sp. Pl. i, 169 (1753).

No corolliferous flowers were seen. Whitewood, altitude 1,100 m., July 7; Custer, altitude 1,650 m., July 15 (No. 851).

Campanula rotundifolia L. Sp. Pl. i, 163 (1753).

Hot Springs, altitude 1,100 m., June 17; Hermosa, altitude 1,150 m., June 22; Lead City, altitude 1,700 m., July 1 (No. 852).

Campanula aparinoides Pursh. Fl. i, 159 (1811).

Wet meadows: Custer, altitude 1,600 m., August 19 (No. 853).

ERICACEÆ.

Vaccinium myrtillus microphyllum Hook. Fl. Bor. Amer. ii, 33 (1834).

The plant (without flowers or fruit) was collected on a shaded hillside, near Lead City, altitude 1,600 m., July 6 (No. 854).

Arctostaphylos uva-ursi (L.) Spreng. Syst. ii, 287 (1825); *Arbutus uva-ursi* L. Sp. Pl. i, 395 (1753).

Common throughout the Black Hills and generally called "kinnikinick." In woods: Custer, altitude 1,700 m., June 3 (No. 855).

PYROLACEÆ.

Pyrola secunda L. Sp. Pl. i, 396 (1753).

In woods: Little Elk, altitude 1,100 m., June 27; Lead City, altitude 1,600 m., July 6 (No. 856).

Pyrola chlorantha Swartz. Kongl. Vet. Acad. Handl. ser. 2, xxxi, 191 (1810).

Woods: Little Elk Canyon, altitude 1,100 m., June 27; Custer, altitude 1,600 m., August 19 (No. 857).

Pyrola elliptica Nutt. Gen. i, 273 (1818).

Shaded hillsides: Lead City, altitude 1,600 m., July 6; Custer, altitude 1,600 m., August 19 (No. 858).

Pyrola rotundifolia L. Sp. Pl. i, 396 (1753).

Only one specimen, collected at Lead City, altitude 1,600 m., July 6 (No. 859).

Pyrola rotundifolia bracteata (Nutt.) Gray. Bot. Cal. i, 160 (1876); *Pyrola bracteata* Nutt. Gen. i, 270 (1818).

The large bracts and purplish flowers distinguish this variety, which is reputed to be of a more westerly range. In a cold bog above Sylvan Lake, at an altitude of 2,100 m., July 19 (No. 860).

Pterospora andromedea Nutt. Gen. i, 269 (1818).

In fruit from preceding year: Custer, altitude 1,650 m., June 3; in flower, Sylvan Lake, altitude 1,200 m., August 19 (No. 861).

PRIMULACEÆ.

Dodecatheon pauciflorum (Durand) Greene, Pittonia. ii, 72 (1889); *Dodecatheon meadia pauciflorum* Durand, Pl. Pratt. 95 (1855).

On wooded hillsides: Custer, altitude 1,700 m., May 28 and June 2 (No. 862).

A stouter form, approaching *D. meadia*, was collected at Custer, June 5 (No. 863).

Androsace septentrionalis L. Sp. Pl. i, 142 (1753).

Custer, altitude 1,650 m., June 3; Elk Canyon, altitude 1,200 m., June 29 (No. 864).

Androsace occidentalis Pursh. Fl. i, 137 (1811).

Dry table-land: Hot Springs, altitude 1,100 m., June 13 (No. 865).

Steironema ciliatum (L.) Bando, Ann. Sci. Nat. ser. 2, xx, 346 (1843); *Lysimachia ciliata* L. Sp. Pl. 117 (1753).

The leaves of the specimens from Custer are scarcely subcordate at the base. Rochford, altitude 1,600 m., July 11; east of Custer, on the French Creek, altitude, 1,500 m., July 22 (No. 866).

Naumburgia thyrsiflora (L.) Duby, in DC. Prodr. viii, 60 (1844); *Lysimachia thyrsiflora* L. Sp. Pl. i, 147 (1753).

Wet places: Elk Canyon, altitude 1,200 m., July 29 (No. 867).

Centunculus minimus L. Sp. Pl. i, 116 (1753).

In Ruby Gulch, northwest of Custer, altitude 1,700 m , August 20 (No. 868).

OLEACEÆ.

Fraxinus pennsylvanica Marsh. Arb. Amer. 51 (1785).

Along Squaw Creek, above Hermosa, altitude 1,050 m., June 22 (No. 869).

Fraxinus pennsylvanica lanceolata (Borck.) Sargent, Silva Amer. vi, 50 (1894); *Fraxinus lanceolata* Borck. Handb. Forstbot. i, 826 (1800).

Together with the preceding (No. 870).

APOCYNACEÆ.

Apocynum androsæmifolium L. Sp. Pl. i, 213 (1753).

Borders of woods: Hermosa, altitude 1,100 m., June 24; Rapid City, altitude 1,000 m., July 25; Little Elk Canyon, altitude 1,100 m., June 27; Lead City, altitude 1,700 m., July 6 (No. 871). Specimens collected in Elk Canyon, altitude 1,200 m., June 29, and at Hot Springs, altitude 1,100 m., August 3, are ambiguous between this species and *A. cannabinum*. The leaves are short-petioled, and the corolla greenish white and smaller than in *A. androsæmifolium*, and the branches upright (No. 872).

ASCLEPIADACEÆ.

Asclepias ovalifolia Dec. in DC. Prodr. viii, 567 (1844).

On French Creek, below Custer, altitude 1,500 m., July 22 (No. 873).

Asclepias verticillata pumila Gray, Proc. Amer. Acad. xii. 71 (1876).

Prairie: Minnekahta, altitude 1,300 m., August 6 (No. 874).

Acerates angustifolia[1] (Nutt.) Dec. in DC. Prodr. viii, 522 (1844); *Polyotus angustifolius* Nutt. Trans. Amer. Phil. Soc. n. ser. v, 20 (1833-1837).

Minnekahta Plains, altitude 1,300 m., August 6 (No. 875).

Acerates viridiflora[2] (Raf.) Eat. Man. ed. 5, 90 (1829); *Asclepias viridiflora* Raf. Med. Rep. xi, 360 (1808).

Only one specimen was collected: Minnekahta Plains, altitude 1,300 m., August 6 (No. 876).

GENTIANACEÆ.

Gentiana acuta Mx. Fl. i, 177 (1803); *Gentiana amarella acuta* Herder, Act. Hort. Petrop. i, 128 (1872).

The specimen appears to be a form of this species modified by the habitat, viz, a heavily shaded hillside. The whole plant is light green, the leaves broad and thin, 3- to 5-ribbed. Young plants can scarcely be distinguished from those of the next. The flowers are greenish yellow, occasionally a little bluish on the limb. South of Custer, altitude 1,700 m., August 15 (No. 877).

[1] See my remarks on this species, this volume, p. 169.

[2] The binomial is not given by Elliott (Bot. S. Car. & Georg. 317, 1823), but he refers *Asclepias viridiflora* to the genus *Acerates*.

Tetragonanthus deflexus (Smith) Kuntze, Rev. Gen. Pl. ii, 431 (1891); *Swertia deflexa* Smith, Rees's Cycl. no. 8 (1816).

In woods: Deadwood, altitude 1,400 m., July 5; Custer, altitude 1,700 m., August 15 (No. 878).

Frasera speciosa Dougl.; Hook. Fl. Bor. Amer. ii, 66 (1838).

High barren hills: Lead City, altitude 1,700 m., July 2; Limestone District, altitude 1,900 m., July 25 (No. 879).

POLEMONIACEÆ.

Phlox douglasii Hook. Fl. Bor. Amer. ii, 73 (1838).

Common around Hot Springs, altitude 1,100 m., June 15 (No. 881).

On wooded hillsides was found a peculiar Phlox, probably belonging to this species. The plant is prostrate, spreading, slender, with very narrow, subulate, weak, leaves, the corolla scarcely longer than the calyx. Hot Springs, altitude 1,300 m., June 15 (No. 880).

Phlox douglasii andicola (Nutt.) Britton, Mem. Torr. Club. v, 269 (1895); *Phlox andicola* Nutt.; Gray, Proc. Amer. Acad. viii, 251 (1870).

Table-land: Hot Springs, altitude 1,100 m., June 14 (No. 882).

Phlox kelseyi Britton, Bull. Torr. Club, xix, 225 (1892).

All specimens collected differ from the typical *Phlox kelseyi* in having nearly white instead of blue or lilac flowers. Two forms were found. One of them has long leaves 1.5 to 3 cm. long and 2 to 3 mm. wide, and larger flowers on pedicels 1 to 2 cm. long. Hot Springs, altitude 1,100 m., June 13; Sylvan Lake, altitude 2,000 m., July 19 (No. 883). The other is more cespitose, has shorter and broader leaves, 0.5 to 1.5 cm. long and 3 to 4 mm. wide, and the nearly sessile flowers have shorter tubes. This is evidently the same as *P. longifolia brevifolia* Gray,[1] collected by Jenney, although I have not seen that plant. My specimens were collected about 20 miles from Jenneys Stockade, in the Limestone District, near Bull Springs, altitude 1,900 m., July 26. Britton refers Jenney's plant to *Phlox kelseyi*. The original variety *brevifolia*[2] is to be referred to *Phlox stansburyi* (Torr.) Britton. From this *Phlox kelseyi* is easily distinguished by its larger flowers and the very broad, obovate lobes of the corolla (No. 884).

Collomia linearis Nutt. Gen. i, 126 (1818).

Hermosa, altitude 1,100 m., June 23; Rochford, altitude 1,700 m., July 11; Lead City, altitude 1,500 m., July 17. The specimens from the latter place are unusually large and branching, some even 4 to 5 dm. high (No. 885).

Gilia spicata capitata Gray, Proc. Amer. Acad. viii, 274 (1870).

I believe that this is distinct from *G. spicata*. It resembles it in foliage, but the form of the corolla is different. Dry hills or table-lands: Hot Springs, altitude 1,100 m., June 13; Deadwood, altitude 1,500 m., July 5; Bull Springs, altitude 1,900 m., July 25 (No. 886).

HYDROPHYLLACEÆ.

Macrocalyx nyctelea (L.) Kuntze, Rev. Gen. Pl. ii, 434 (1891); *Ipomœa nyctelea* L. Sp. Pl. i, 160 (1753).

Edgemont, altitude 1,050 m., May 27; Rochford, altitude 1,600 m., July 11; Oreville, altitude 1,630 m., July 16 (No. 887).

BORAGINACEÆ.

Lappula virginiana (L.) Greene, Pittonia, ii, 182 (1891); *Myosotis virginica* L. Sp. Pl. i, 131 (1753).

Only one specimen in bloom, found near Lead City, altitude 1,600 m., July 6 (No. 888).

[1] In Newton & Jenney, Geol. Surv. Black Hills, 535 (1880).
[2] Gray, Syn. Fl. ii, pt. i, 133 (1878).

Lappula deflexa americana (Gray) Greene, Pittonia, ii, 183 (1891); *Echinospermum deflexum americanum* Gray, Proc. Amer. Acad. xvii, 224 (1882).

My specimens belong to a form with short, broad leaves and small flowers. Rare: Lead City, altitude 1.600 m., July 6 (No. 889).

Lappula floribunda (Lehm.) Greene, Pittonia, ii, 182 (1891); *Echinospermum floribundum* Lehm. Pug. ii, 24 (1830).

The common form has distinctly pinnately nerved leaves. Hot Springs, altitude 1,100 m., June 13; Elk Canyon, altitude 1,200 m., June 29; Custer, altitude 1,700 m., July 16 (No. 890).

At Oreville, altitude 1,650 m., July 16, a form was collected which differs, in having much thicker, narrowly lanceolate leaves, with a prominent midrib but obsolete lateral veins (No. 891).

At Rochford, altitude 1,650 m., July 11, a plant was collected, which I also refer here, although it differs considerably from the common form. It was collected in flower only, and as the fruit is necessary for full identification, I leave it with this species. The plant is tall, 1 to 2 m. high, branched from an apparently perennial root, making big, bushy clumps of a dozen stems or more. The upper parts of the plant are yellowish silky, the lower somewhat strigose; the lower leaves spatulate, the upper lanceolate, thickish, with a prominent midrib, the lateral veins obsolete (No. 892).

Lappula redowskii occidentalis (Wats.) Rydberg, Contr. Nat. Herb. iii, 170 (1895); *Echinospermum redowskii occidentale* (Wats.) Bot. King Surv. 246 (1871).

Some of the specimens resemble *L. lappula* in having a larger, more campanulate corolla and being more branched from the base and more leafy. The immature fruit shows characters which place it with *L. redowskii*. Edgemont, altitude 1.050 m., May 27; Hot Springs, altitude 1,100 m., June 15 (No. 576).

The common form was collected at Hot Springs, altitude 1,100 m., June 13; Hermosa, altitude 1,050 m., June 23; Custer, altitude 1,700 m., July 16 (No. 577).

Oreocarya glomerata (Pursh) Greene, Pittonia, i, 58 (1887); *Cynoglossum glomeratum* Pursh. Fl. ii, 729 (1814).

High table-lands and hills: Hot Springs, altitude 1,100 m., June 13; Whitewood, altitude 1,100 m., July 7 (No. 893).

Cryptanthe pattersoni (Gray) Greene, Pittonia, i, 120 (1887); *Krynitzkia pattersoni* Gray, Proc. Amer. Acad. xx, 268 (1885).

The seeds are brown-spotted and less attenuate than in *C. fendleri*, but otherwise as in that species. The leaves are also broader. Lead City, altitude 1,600 m., July 6; Rochford, altitude 1,700 m., July 12 (No. 894).

No. 895 is a small Cryptanthe, perhaps nearly related, but there are no fruits in the collection, hence it can not be determined. It looks very like some specimens in the National Herbarium labeled *Krynitzkia affinis*, but it may be an undeveloped form of nearly any of the related species. Buckhorn Mountain, near Custer, altitude 1,800 m., July 16.

Myosotis verna macrosperma (Engelm.); *Myosotis macrosperma* Engelm. Amer. Journ. Sci. xlvi, 98 (1844).

Rare: Hot Springs, altitude 1,100 m., June 13 (No. 575).

Myosotis sylvatica Hoffmann, Deutsch. Fl. i, 85 (1791).

Slender, 1.5 to 3 dm. high, raceme loose, the pedicels longer than the fruiting calyx. It does not belong to the variety *alpestris*, which has been regarded as the only American form, but rather to the species. High altitudes in damp places among rocks: Sylvan Lake, altitude 2,000 m., July 19 (No. 896).

Mertensia sibirica (L.) Don, Hist. Dichl. Pl. iv, 319 (1838); *Pulmonaria sibirica* L. Sp. Pl. i, 135 (1753).

A single fruiting specimen, which seems to belong to this species, was collected at Rochford, altitude 1,700 m., July 12 (No. 897).

Mertensia paniculata (Ait.) Don, Hist. Diehl. Pl. iv, 318 (1838); *Pulmonaria paniculata* Ait. Hort. Kew. i, 181 (1789).

In the few specimens collected the calyx is not ciliate and the corolla only 6 to 8 mm. long. Rochford, altitude 1,700 m., July 12 (No. 898).

Mertensia lanceolata (Pursh) DC. Prodr. x, 88 (1816); *Pulmonaria lanceolata* Pursh, Fl. ii, 729 (1814).

There seem to be two forms of this species, one with larger flowers, about 1 cm. long, the tube more abruptly widening into the campanulate limb, thicker, somewhat fleshy leaves, and more simple stem. This is the more common form in the Black Hills. Custer, altitude 1,700 m., May 30 (No. 899).

The form growing in western Nebraska with thin leaves of a light-green color, paniculately branched stem, and smaller, more funnelform flowers was found near Sylvan Lake, altitude 2,000 m., July 19 (No. 900).

Lithospermum angustifolium Mx. Fl. i, 130 (1803).

Prairie: Hot Springs, altitude 1,100 m., June 15; Buffalo Gap, altitude 1,100 m., June 21; Custer, altitude 1,650 m., August 1 (No. 901).

Onosmodium molle Mx. Fl. i, 133 (1803).

Prairie: Hot Springs, altitude 1,100 m., June 16 (No. 902).

CONVOLVULACEÆ.

Evolvulus nuttallianus Roem. & Schult. Syst. Veg. vi, 198 (1820); *E. argenteus* Pursh, Fl. i, 187 (1814), not R. Br. Prodr. (1810); *E. pilosus* Nutt. Gen. i, 174 (1818), not Lam.

Rare: collected in fruit only at Hot Springs, altitude 1,100 m., June 13 (No. 903).

Ipomœa leptophylla Torr. in Frem. First Rep. 94 (1843).

Prairie: Hot Springs, altitude 1,100 m., June 15 (No. 578).

Convolvulus sepium L. Sp. Pl. i, 153 (1753).

Rare: Custer, altitude 1,650 m., August 1 (No. 904).

SOLANACEÆ.

Solanum triflorum Nutt. Gen. i, 128 (1818).

On the railroad embankment north of Custer, altitude 1,650 m., July 16 (No. 905).

Solanum nigrum L. Sp. Pl. i, 186 (1753).

Hot Springs, altitude 1,050 m., June 13 (No. 906).

Solanum rostratum Dunal, Hist. Sol. 231 (1813).

Custer, altitude 1,650 m., August 1 (No. 907).

Physalis heterophylla Nees, Linnæa, vi, 463 (1831); *Physalis viscosa* Pursh. Fl. i, 157 (1814), not L.; *Physalis virginiana* Gray, Proc. Amer. Acad. x, 65 (1871), not Mill.

This is an upright form with thinner leaves and scarcely glandular at all. Hills, on French Creek east of Custer, altitude 1,500 m., July 23 (No. 908).

Physalis longifolia Nutt. Trans. Amer. Phil. Soc. ser. 2, v, 193 (1857).

Among bushes: Hot Springs, altitude 1,050 m., June 18 (No. 910).

Physalis virginiana Mill. Gard. Dict. ed. 8, no. 4 (1768); *Physalis lanceolata* Gray, Proc. Amer. Acad. x, 67 (1871), not Mx.

This is the common form of *P. virginiana*. It differs from the type slightly in the leaves, which are less sinuately toothed. The original *P. virginiana*, described and figured by Miller, has deeply toothed leaves and the whole plant is more or less glutinous. It is a very rare form.

The few poor specimens in this collection are more or less pubescent, with sinuately toothed or wavy-margined leaves, yellow fruit, and a pyramidal, angled, fruiting calyx with a sunken base, a character which distinguishes all forms of this species from *P. lanceolata* Mx. Lead City, altitude 1,600 m., July 6 (No. 909).

SCROPHULARIACEÆ.

Verbascum thapsus L. Sp. Pl. i, 177 (1753).
Introduced on the railroad embankment near Fall River Falls, altitude 1,000 m., August 10 (No. 911). Only two plants collected.

Collinsia parviflora Lindl. Bot. Reg. xiii, t. 1082 (1827).
Dry hillsides: Little Elk Canyon, altitude 1,200 m., June 27; Elk Canyon, altitude 1,200 m., June 29 (No. 913).

Linaria canadensis (L.) Dum. Bot. Cult. ii, 96 (1802); *Antirrhinum canadense* L. Sp. Pl. ii, 618 (1753).
Very slender and depauperate, apparently with cleistogamous flowers. The same form has also been collected in Nebraska by Rev. J. M. Bates, of Valentine. Custer, altitude 1,700 m., August 20 (No. 912).

Scrophularia nodosa occidentalis, var. nov.
Tall, 1 to 1.5 m. high, glandular, especially on the upper part of the stem; leaves ovate or slightly heartshaped at the base, doubly and sharply serrate or incised; petioled, with fascicles of smaller leaves in the axils; panicle with short branches; sepals rounded-elliptical, obtuse, slightly margined; corolla lurid-greenish, gibbose at the base; sterile stamens very broad, kidney-shaped on a claw.
It differs from *S. nodosa* proper and *S. nodosa marilandica* in being glandular and in the sharp serration of the leaves; from *S. californica* in its larger flowers, sharper serrations, stout habit, and the form of the sterile stamen; from all three in the more gibbose corolla. No. 997, Suksdorf seems to belong to the same variety. Rapid City, altitude 1,000 m., July 25 (No. 911). Most of the specimens were damaged by rain while in the press.

Pentstemon grandiflorus Nutt. Fraser's Cat. (1813).
Prairies: Hermosa, altitude 1,050 m., June 21 (No. 915).

Pentstemon glaber Pursh, Fl. ii, 738 (1814).
Hills: North of Deadwood, altitude 1,500 m., July 5; Rochford, altitude 1,700 m., July 12; Custer, altitude 1,700 m., July 18 (No. 916).

Pentstemon angustifolius Pursh, Fl. ii, 738 (1814), not Lindl. (1827); *P. cærulens* Nutt. Gen. ii, 52 (1818).
Only two specimens collected: Hot Springs, altitude 1,100 m., June 15 (No. 917).

Pentstemon jamesii Benth. in DC. Prodr. x, 325 (1846).
My specimens are like Fendler's No. 575. This, with his No. 579 and the original specimens of James, are the only ones cited by Gray in his synopsis of the genus in the Proceedings of the American Academy.[1] I think Watson's No. 778, named *P. cristatus*, should also be referred to this species. Table-land: Hot Springs, altitude 1,100 m., June 16 (No. 918).

Pentstemon erianthera Pursh, Fl. ii, 737 (1814).
P. cristatus Nutt.[2] is a *nomen nudum*; hence *P. erianthera* Pursh, is the oldest name. Only four specimens of this were collected, two in Elk Canyon, altitude 1,200 m , June 29, and two depauperate ones in the Limestone District near Bull Springs, altitude 1,900 m., July 27 (No. 919).

Pentstemon albidus Nutt. Gen. ii, 53 (1818).
Only one specimen, found near Hermosa, altitude 1,100 m., June 23 (No. 920).

Pentstemon gracilis Nutt. Gen. ii, 52 (1818).
Common: Hot Springs, altitude 1,050 m., June 15; Hermosa, altitude 1,000 m., June 22; Elk Canyon, altitude 1,200 m., June 29; Lead City, altitude 1,600 m., July 6; Rochford, altitude 1,700 m., July 11 (No. 921).

Mimulus luteus L. Sp. Pl. ed. 2, ii, 884 (1763).
In a wet, shady place near a stream, southwest of Lead City, altitude 1,700 m., July 9 (No. 922).

---

[1] vi, 67 (1866).   [2] Fraser's Cat. (1813).

Mimulus glabratus jamesii ,Torr. & Gray) Gray, Syn. Fl. Suppl. 417 (1886);
Mimulus jamesii Torr. & Gr.; DC. Prodr. x, 371 (1846).
In water: Hot Springs, altitude 1,050 m., June 18; Custer, altitude 1,600 m.,
August 1; Pringle, altitude 1,500 m., August 5 (No.923).

Gratiola virginiana L. Sp. Pl. i, 17 (1753).
The corolla in my specimens is fully twice as long as the calyx. In marshy places:
Buckhorn Mountain, near Custer, altitude 1,709 m., July 16; west of Custer, alti-
tude 1,700 m., July 25 (No.921).

Wulfenia rubra (Hook.) Greene, Erythea, ii, 83 (1891); *Gymnandra rubra* Hook.
Fl. Bor. Amer. ii, 103 (1838).
Hillsides: Custer, altitude 1,100 m., June 6 (No.925).

Veronica anagallis L. Sp. Pl. i, 12 (1753).
Rare: Hot Springs, altitude 1,050 m., June 16 (No.926).

Veronica americana Schwein.; Benth. in DC. Prodr. x, 468 (1846).
Rapid Creek, altitude 1,000 m., June 25; Whitewood, altitude 1,150 m., July 7;
Custer, altitude 1,600 m., August 1 (No.927).

Veronica peregrina L. Sp. Pl. i, 14 (1753).
My specimens are decidedly glandular. Hills: Hermosa, altitude 1,100 m., June
23; Lead City, altitude 1,700 m., July 6 (No.928).

Castilleja acuminata (Pursh) Spreng. Syst. ii, 775 (1825); *Bartsia acuminata*
Pursh, Fl. ii, 429 (1814).
Woods: Little Elk Canyon, altitude 1,100 m., June 27; Elk Canyon, altitude
1,300 m., June 29; Lead City, altitude 1,600 m., July 6 (No. 929).

Castilleja sessiliflora Pursh, Fl. ii, 738 (1814).
Table-land: Hot Springs, altitude 1,100 m., June 15 (No. 930).

Orthocarpus luteus Nutt. Gen. ii, 57 (1818).
Meadow: Custer, altitude 1,600 m., August 1 (No. 931).

VERBENACEÆ.

Verbena stricta Vent. Hort. Cels. t. 53 (1800).
Custer, altitude 1,600 m., August 1; Hot Springs, altitude 1,100 m., August 3
(No. 932).

Verbena hastata L. Sp. Pl. i, 20 (1753).
Rare: Hot Springs, altitude 1,050 m., June 16 (No. 933).

Verbena bracteosa Mx. Fl. ii, 13 (1803).
Lead City, altitude 1,600 m., July 9 (No. 934).

Verbena bipinnatifida Nutt. Journ. Acad. Phila. ii, 123 (1821).
Prairie, 2 miles east of Fall River Falls, altitude 1,000 m., June 18 (No. 935).

OROBANCHACEÆ.

Thalesia fasciculata (Nutt.) Britton, Mem. Torr. Club, v, 298 (1894); *Orobanche
fasciculata* Nutt. Gen. ii, 59 (1818).
Rare: Custer, altitude 1,650 m., June 6 (No. 936).

Orobanche ludoviciana Nutt. Gen. ii, 58 (1818).
On the railroad embankment between Hot Springs and Fall River Falls, altitude
1,050 m., August 8 (No. 937).

LABIATÆ.

Mentha canadensis glabrata Benth. Lab. 181 (1833).
Gray, in the Synoptical Flora, cites *M. borealis* Mx. as a synonym, but this plant
does not agree with the description of Michaux's species, which perhaps is the
typical *M. canadensis* and not the variety.
Hot Springs, altitude 1,050 m., August 3; Custer, altitude 1,650 m., August 20
(No. 938).

Lycopus sinuatus Ell. Bot. S. Car. & Georg. i, 26 (1816).
Wet meadows: Custer, altitude 1,650 m., August 20 (No. 939).

Hedeoma hispida Pursh, Fl. ii, 414 (1814).
Dry places: Rochford, altitude 1,700, July 11; Custer, altitude 1,700 m., August 1; Minnekahta, altitude 1,300 m., August 1 (No. 940).

Hedeoma drummondii Benth. Lab. 368 (1834).
Dry soil: Custer, altitude 1,700 m., July 16; Hot Springs, altitude 1,100 m., August 2 (No. 941).

Salvia lanceolata Willd. Enum. 37 (1809).
The leaves are broadly oblong or spatulate. Custer, altitude 1,700 m., July 16; Hot Springs, altitude 1,100 m., August 2 (No. 942).

Monarda fistulosa mollis (L.) Benth. Lab. 317 (1833); *Monarda mollis* L. Amœn. Acad. iii, 399 (1756).
Hot Springs, altitude 1,100 m., August 2; Custer, altitude 1,650 m., August 15 (No. 943); Rochford, altitude 1,700 m., July 12 (No. 944).

Vleckia anethiodora (Nutt.) Greene, Mem. Torr. Club, v, 282 (1894); *Hyssopus anethiodorus* Nutt. Fraser's Cat. (1813).
Among bushes: Whitewood, altitude 1,100 m., July 7; east of Custer, on the French Creek, altitude 1,600 m., July 22 (No. 945).

Prunella vulgaris L. Sp. Pl. ii, 600 (1753).
In damp woods: Whitewood, altitude 1,150 m., July 7; east of Custer, altitude 1,600 m., July 22 (No. 946).

Dracocephalum parviflorum Nutt. Gen. ii, 35 (1818).
Hills: Elk Canyon, altitude 1,300 m., June 29; Rochford, altitude 1,800 m., July 12 (No. 947).

Scutellaria galericulata L. Sp. Pl. ii, 599 (1753).
In French Creek, near Custer, altitude 1,600 m., July 16 and August 1 (No. 948).

Stachys palustris L. Sp. Pl. ii, 580 (1753).
Two forms were met with, one with short, oblong leaves, smaller and darker flowers, and more hairy stem (No. 949); the other with longer, lanceolate leaves, and larger light-colored flowers (No. 1208). Lead City, altitude 1,700 m., July 7; Custer, altitude 1,700 m., August 20.

Stachys aspera Mx. Fl. ii, 5 (1803).
Custer, altitude 1,700 m., August 20 (No. 950).

PLANTAGINACEÆ.

Plantago major L. Sp. Pl. i, 112 (1753).
Rochford, altitude 1,600 m., July 11; Custer, altitude 1,650 m., August 1 (No. 951).

Plantago purshii Roem. & Schult. Syst. iii, 120 (1818).
Dry plains: Minnekahta, altitude 1,300 m., August 5 (No. 952).

NYCTAGINACEÆ.

Allionia nyctaginea Mx. Fl. i, 100 (1803).
Rare: Hot Springs, altitude 1,050 m., August 2 (No. 953).

Allionia hirsuta Pursh, Fl. ii, 728 (1814).
Two forms were met with, which seem very distinct, but my collection from the Sand Hills of central Nebraska shows that they grade into each other. One has very broad, oblong-lanceolate, or ovate-oblong leaves, the stem hairy throughout. Custer, altitude 1,700 m., August 1 (No. 954). The other has narrow, lanceolate leaves, the stem hairy only at the nodes. Rochford, altitude 1,700 m., July 11; Hot Springs, altitude 1,100 m., August 2 (No. 955).

Allionia albida Walt. Fl. Car. 81 (1788).

There are two forms in the collection, which I refer to this species. The material is too scanty and poor for a satisfactory determination. One form with broadly lanceolate leaves was obtained at Hot Springs, altitude 1,100 m., August 3 (No. 956). The other form, with very narrowly lanceolate or linear leaves, grew on the plains between Custer and Fairburn, altitude 1,100 m., July 23 (No. 957). Both forms differ from *A. hirsuta* in being smooth up to the peduncle, and from *A. linearis* in the leaves, which are undulate and sparingly ciliolate on the margin, and in the peduncles and involucres, which are hispid.

Allionia linearis Pursh, Fl. ii, 728 (1814).

Custer, altitude 1,650 m., August 1; Hot Springs, altitude 1,100 m., August 3 (No. 958).

AMARANTHACEÆ

Amaranthus blitoides Wats. Proc. Amer. Acad. xii, 273 (1877).

Railroad embankment, Minnekahta, altitude 1,270 m., August 1 (No. 960).

CHENOPODIACEÆ.

Chenopodium hybridum L. Sp. Pl. i, 219 (1753).

One specimen, collected at Hot Springs, altitude 1,050 m., June 17 (No. 961).

Chenopodium album L. Sp. Pl. i, 219 (1753).

Hot Springs, altitude 1,050 m., June 17 (No. 962).

Chenopodium fremonti Wats. Bot. King Surv. 287 (1871).

In damp woods: Sylvan Lake, altitude 1,900 m., July 20 (No. 963).

Chenopodium fremonti incanum Wats. Proc. Amer. Acad. ix, 91 (1871).

Two small specimens collected near Fall River Falls, altitude 1,000 m., June 17 (No. 964).

Chenopodium leptophyllum (Moq.) Nutt.; Moq. in DC. Prodr. xiii, pt. 2, 71 (1849), as synonym; *C. album leptophyllum* Moq. in DC. Prodr. loc. cit.

Hot Springs, altitude 1,100 m., June 17 (No. 965).

Chenopodium capitatum (L.) Wats. Bot. Cal. ii, 18 (1880); *Blitum capitatum* L. Sp. Pl. i, 4 (1753).

Near the railroad, at Rochford, altitude 1,600 m., July 11 (No. 966).

Monolepis nuttalliana (Roem. & Schult.) Greene, Fl. Fran. 168 (1891); *Blitum nuttallianum* Roem. & Schult. Syst. Mant. i, 65 (1822).

The leaves are more or less sinuately lobed, and the stem more upright than in Nebraska specimens. Buffalo Gap, altitude 1,000 m., June 21; Rochford, altitude 1,600 m., July 11 (No. 967).

POLYGONACEÆ.

Eriogonum flavum Nutt. Fraser's Cat. 1813.

Dry hills: Hermosa, altitude 1,100 m., June 23; Lead City, altitude 1,700 m., July 4 (No. 968).

Eriogonum annuum Nutt. Trans. Amer. Phil. Soc. ser. 2, v, 164 (1833–1837).

Hot Springs, altitude 1,100 m., August 3 (No. 969).

Eriogonum pauciflorum Pursh, Fl. ii, 735 (1814).

The specimens agree well with the description of this species, except that the involucre is narrowly turbinate, and the lobes scarcely scarious-margined, and that the scape is more slender and the leaves are nearly glabrous above. Dry hills: Hermosa, altitude 1,100 m., June 23 (No. 970).

Eriogonum multiceps Nees, in Max. Reise N. A. ii, 116 (1841).

Gypsum hills, near Hot Springs, altitude 1,100 m., August 3 (No. 971).

Rumex venosus Pursh, Fl. ii, 733 (1814).

Custer, altitude 1.700 m., June 5; Hermosa, altitude 1,100 m., June 27 (No. 972).

Rumex occidentalis Wats. Proc. Amer. Acad. xii, 253 (1876).

Near water: Buffalo Gap, altitude 975 m., June 21; Custer, altitude 1,600 m., August 1 (No. 973).

Rumex salicifolius Weinm. Fl. iv, 28 (1821).

Rochford, altitude 1,600 m., July 12 (No. 974).

Rumex crispus L. Sp. Pl. i, 335 (1753).

I believe this is a native of the Black Hills as well as of western Nebraska. Lead City, altitude 1,600 m., July 9; Custer, altitude 1,600 m., August 1 (No. 975).

Rumex acetosella L. Sp. Pl. i, 338 (1753).

Introduced: near Whitewood, altitude 1,100 m., July 7 (No. 976).

Polygonum aviculare L. Sp. Pl. i, 362 (1753).

Rochford, altitude 1,600 m., July 11 (No. 977).

No. 978 is a small, undeveloped, erect form which I took to belong to this species. Mr. J. K. Small, who has identified the species of Polygonum, writes: "It might be a form of *aviculare, litorale,* or *ramosissimum.*"

Polygonum litorale Link, Schrad. Journ. Bot. i, 54 (1799).

Common along the railroad above Hot Springs, altitude 1,050 m., August 9 (No. 979). Some specimens are more upright and have broad, elliptical leaves (No. 980).

Polygonum ramosissimum Mx. Fl. i, 237 (1803).

Hot Springs. altitude 1,050 m., August 9 (No. 982).

Some luxuriant specimens, more decumbent and spreading, with thicker leaves, resemble, according to Mr. Small, a form that has been named variety *patulum* by Engelmann in manuscript. Hot Springs. August 9 (No. 981).

Polygonum sawatchense Small, Bull. Torr. Club, xx, 213 (1893).

Mr. Small remarks about this: "Slender form due to the lower altitude than that of the type." Custer, altitude 1,800 m., July 16 (No. 983).

Polygonum douglasii Greene, Bull. Cal. Acad. i, 125 (1885).

Tall, 4 to 7 dm. high, with conspicuous, strongly nerved sheath. Custer, altitude 1,700 m., August 1 (No. 984).

Polygonum lapathifolium L. Sp. Pl. i, 360 (1753).

Hot Springs, altitude 1,050 m., August 9 (No. 985).

Polygonum emersum (Mx.) Britton, Trans. N. Y. Acad. viii, 73 (1889); *Polygonum amphibium emersum* Mx. Fl. i, 240 (1803).

French Creek, east of Custer, altitude 1,100 m., July 22 (No. 986).

Polygonum viviparum L. Sp. Pl. i, 360 (1753).

In wet place, among moss, below Sylvan Lake, altitude 2,000 m., July 19 (No. 987).

Polygonum convolvulus L. Sp. Pl. i, 364 (1753).

Railroad embankment, Hot Springs, altitude 1,050 m., August 9 (No. 988).

ELEAGNACEÆ.

Eleagnus argentea Pursh, Fl. i, 114 (1814).

Hills in the Limestone District, near Bull Springs, altitude 1,900 m., July 27, (No. 989).

Lepargyræa canadensis (L.) Greene, Pittonia, ii, 122 (1890); *Hippophaë canadensis* L. Sp. Pl. ii, 1024 (1753).

Woods: Custer, altitude 1,700 m., May 30; Little Elk, altitude 1,100 m., June 28; Lead City, altitude 1,600 m., July 6 (No. 990).

Lepargyræa argentea (Nutt.) Greene, Pittonia, ii, 122 (1890); *Eleagnus argentea* Nutt. Fraser's Cat. (1813).

A few bushes were seen near Hot Springs, but no specimens were secured.

522

SANTALACEÆ.

Comandra pallida A. DC. Prodr. xiv, 636 (1857).
Table-land: Hot Springs, altitude 1,100 m., June 19; Hermosa, altitude 1,100 m., June 23 (No. 991).

EUPHORBIACEÆ.

Euphorbia glyptosperma Engelm. Bot. Mex. Bound. 187 (1859).
Sandy soil: Hot Springs, altitude 1,050 m., August 9 (No. 992).

Euphorbia hexagona Nutt.; Spreng. Syst. iii, 791 (1826).
Sand draw: Hot Springs, altitude 1,150 m., August 9 (No. 993).

Euphorbia marginata Pursh, Fl. ii, 607 (1814).
Hot Springs, altitude 1,050 m., August 9 (No. 994).

Euphorbia dentata Mx. Fl. ii, 211 (1803).
A very variable species. In the same spot were found specimens which might be referred to *E. dentata* proper, to variety *rigida*, and to variety *linearis*. Some even approached *E. cuphosperma* Engelm., which seems to me, however, to be a variety of *E. dentata*. In several cases the seeds approach those of *E. cuphosperma*, being more or less ovoid-pyramidal, with a groove on one side. The leaves are from broadly ovate to linear, and from coarsely dentate to nearly entire.
Sandy soil: Hot Springs, altitude 1,050 m., June 19 and August 3 (No. 995).

Euphorbia dictyosperma Fisch. & Mey. Ind. Hort. Petrop. ii, 37 (1835).
Hot Springs with the preceding, June 17 and August 3 (No. 996).

Euphorbia montana robusta Engelm. Bot. Mex. Bound. 192 (1859).
Hillsides: Hot Springs, altitude 1,100 m., June 17 (No. 997).

Croton texensis (Klotzsch) Muell. Arg. in DC. Prodr. xv, pt. 2, 692 (1866); *Hendecandra texensis* Klotzsch, in Erichs. Archiv. i, 252 (1841).
Sandy soil: Hot Springs, altitude 1,050 m., August 9 (No. 998).

URTICACEÆ.

Ulmus americana L. Sp. Pl. i, 226 (1753).
Along Fall River, altitude 1,000 to 1,100 m., rare. August 8 (No. 999).

Celtis occidentalis L. Sp. Pl. ii, 1044 (1753).
Only one shrubby specimen seen, near Hot Springs, altitude 1,050 m., August 8 (No. 1000).

Urtica gracilis Ait. Hort. Kew. iii, 341 (1879).
Not common: Custer, altitude 1,650 m., August 1 (No. 1001).

Parietaria pennsylvanica Muhl.; Willd. Sp. Pl. iv, 955 (1805).
In shady woods, rare: Hot Springs, altitude 1,050 m., June 13 (No. 1002).

Humulus lupulus L. Sp. Pl. ii, 1028 (1753).
Along French Creek, below Custer, rare; altitude 1,500 m., July 22 (No. 1003).

CUPULIFERÆ.

Betula papyrifera Marsh. Arb. Amer. 19 (1785).
Common in the Black Hills proper. Elk Canyon, altitude 1,200 m., June 29 (No. 1004).

Betula occidentalis Hook. Fl. Bor. Amer. ii, 155 [1838].
The common form in the Black Hills is a tree about 10 m. high with leaves about as large as those of the preceding. In bloom: Custer, altitude 1,700 m., June 5; Little Elk Canyon, altitude 1,200 m., June 27 (No. 1005). Another form was found scarcely 1 m. high with smaller, sharply and doubly serrate leaves, more glandular stem, shorter aments with shorter lateral lobes to the bracts. It approaches the next somewhat in habit and general appearance, but evidently belongs to *B. occidentalis*. Near a brook: Oreville, altitude 1,625 m., July 23 (No. 1006).

Betula glandulosa Mx. Fl. Bor. Amer. ii, 180 (1803).
A shrub 1 to 1.5 m. high. Rochford, altitude 1,600 m., July 12 (No. 1007).

Corylus rostrata Ait. Hort. Kew. iii. 364 (1789).
Here and there in the hills, but not common: Elk Canyon, altitude 1,200 m.,
June 29 (No. 1008).

Ostrya virginiana (Mill.) Willd. Sp. Pl. iv. 469 (1805); *Carpinus virginiana* Mill.
Dict. ed. 8, no. 1 (1768).
Among the foothills: Rapid City, altitude 1,000 m., June 25 (No. 1009).

Quercus macrocarpa Mx. Hist. Chen. Amer. ii, 2 (1801).
In the foothills. In most places only shrubby, 3 to 6 m. high, knotty. On the
upper Squaw Creek, east of Custer, there were good-sized trees. Hermosa, altitude
1,100 m., June 24; Elk Canyon, altitude 1,200 m., June 29 (No. 1010).
At the first place specimens occurred which had more narrowly lobed leaves and a
more straight and vigorous growth. Only young trees were seen (No. 1011).

SALICACEÆ.

Salix bebbiana Sargent, Gard. & For. viii, 463 (1895); *Salix rostrata* Richards.
App. Frankl. Journ. 753 (1823), not Thuil. Fl. Par. 516 (1790), nor Schleich (1807);
S. vagans occidentalis Anders. Kongl. Akad. Stock. Förh. (1858), not *S. occidentalis*
Bosc. in Koch, Sal. Com. 16 (1828).
A shrub 1 to 3 meters high. Custer, altitude 1,700 m., May 30 (No. 1012); June 6
(No. 1013). In fruit and leaf: Elk Canyon, altitude 1,200 m., June 30 (No. 1016);
Hermosa, altitude 1,100 m., June 24; Custer, altitude 1,700 m., August 16; Hot
Springs, altitude 1,100 m., August 7 (No. 1018). A few specimens with thick leaves
as in *S. humilis* and *S. tristis* were collected at Rochford, altitude 1,700 m., July 12.
Even these have been referred to *S. rostrata* by Mr. Bebb (No. 1019).

Salix discolor Muhl. Neue Schrift. Gesell. Naturf. Freunde Berlin, iv, 234 (1803).
A shrub 2 to 8 m. high. Custer, altitude, 1,700 m., June 1 (No. 1014).

Salix cordata Muhl. Neue Schrift. Gesell. Naturf. Freunde Berlin, iv, 236 (1803).
Near water: Custer, altitude 1,700 m., June 4 (No. 1015); leaves, Lead City, altitude 1,600 m., July 9 (No. 1017).

Salix fluviatilis Nutt. Sylva, 73 (1842); *S. longifolia* Muhl. Neue Serift. Gesell.
Naturf. Freunde Berlin, iv, 236 (1803), not Lam.
Rochford, altitude 1,650 m., July 11 (No. 1020).

Populus tremuloides Mx. Fl. ii, 243 (1803).
Common in the higher parts of the Black Hills. Custer, altitude 1,700 m., June 4,
female flowers (No. 1021).

Populus deltoides Marsh. Arb. Amer. 106 (1785); *P. monilifera* Ait. Hort. Kew.
iii, 106 (1809).
Common in the foothills: Hot Springs, altitude 1,100 m., June 17, fruit: August 3,
leaves (No. 1022).
One tree, with unusually narrow crown, was found in a canyon east of Hot Springs,
August 3. Of this all the leaves were longer than broad, cuneate at the base, with
long acumination and with larger teeth than usual (No. 1023).

Populus acuminata Rydberg, Bull. Torr. Club, xx, 50 (1893).
This is the same as No. 372 of my western Nebraska collection, but the leaves are
broader and with shorter acumination. Only three trees found, near Hot Springs,
altitude 1,100 m., August 3 (No. 1024).

Populus angustifolia James, Long Exped. i, 497 (1823).
Common along Little Elk Creek, altitude 1,100 m., June 28 (No. 1025).

ORCHIDACEÆ.

Corallorhiza corallorhiza (L.) Karst. Deutsch. Fl. 118 (1880-1883); *Ophrys corallorhiza* L., Sp. Pl. ii, 945 (1753).

Little Elk Canyon, altitude 1,100 m., June 27; Lead City, altitude 1,600 m., July 6 (No. 1026).

Corallorhiza multiflora Nutt. Journ. Acad. Phila. iii, 138 (1823).

Elk Canyon, altitude 1,300 m., June 29; Lead City, altitude 1,600 m., July 4; Custer, altitude 1,600 m., July 16; Sylvan Lake, altitude 2,000 m., July 19 (No. 1027).

Habenaria hyperborea (L.) R. Br.; Ait. Hort. Kew. ed. 2, v, 193 (1813); *Orchis hyperborea* L. Mant. 121 (1767).

Wet places: Lead City, altitude 1,600 m., July 6; near Harneys Peak, altitude 2,000 m., July 20 (No. 1028).

Habenaria bracteata (Willd.) R. Br.; Ait. Hort. Kew. ed. 2, v, 192 (1813); *Orchis bracteata* Willd. Sp. Pl. iv, 34 (1805).

Wet places in woods at high altitudes, near Harneys Peak, altitude 2,000 m., July 20 (No. 1029).

Gyrostachys romanzoffiana (Cham.) MacMillan, Metasp. Minn. Val. 174 (1892); *Spiranthes romanzoffiana* Cham. Linnæa, iii, 32 (1828).

Ruby Glen, near Custer, altitude 1,700 m., August 19 (No. 1030).

Peramium repens (L.) Salisb. Trans. Hort. Soc. i, 301 (1812); *Satyrium repens* L. Sp. Pl. ii, 945 (1753).

Only two specimens found: below Sylvan Lake, altitude 1,900 m., July 19 (No. 1031).

Cypripedium parviflorum Salisb. Trans. Linn. Soc. i, 77 (1791).

In woods: Elk Canyon, altitude 1,200 m., June 28 (No. 1032).

IRIDACEÆ.

Iris missouriensis Nutt. Journ. Acad. Phila. vii, 58 (1834).

This is the *I. tolmeiana* Herbert, of Newton and Jenney's Report.

Collected in fruit only: Piedmont, altitude 1,100 m., June 27; Rochford, altitude 1,600 m., July 12; Pringle, altitude 1,500 m., August 5 (No. 1033).

Sisyrinchium bermudiana L. Sp. Pl. ii, 954 (1753).

Among the foothills: Hot Springs, altitude 1,100 m., June 19; Elk Canyon, altitude 1,200 m., June 29 (No. 1034).

LILIACEÆ.

Allium cernuum Roth, in Roem. Arch. Bot. i, pt. 3, 40 (1798).

All specimens of this species, in the National Herbarium, from the Rocky Mountain region have narrow and apparently channeled leaves; those from the eastern United States have broad and flattened leaves. I do not know which form should be regarded as the typical one, as I have not seen the original description. The description and figure in Curtis's Botanical Magazine agree with the specimens of this collection and, as far as I can judge, with all from the Rocky Mountain region. There the leaves are represented as half-round and channeled, not as "sharply keeled" as they are described in Gray's Manual, ed. 6. In mine they are not keeled at all.

Bull Springs, altitude 1,900 m., July 26; Custer, altitude 1,600 m., August 1; Hot Springs, altitude 1,100 m., August 3 (No. 1035).

Allium geyeri Wats. Proc. Amer. Acad. xiv, 227 (1879).

This, as also the next, is described as having crested capsules. The crests are, however, easily overlooked, being 2 small lobes on each valve, near the top. Overlooking these, I named this *A. mutabile* and the next *A. nuttallii*, which they resemble, respectively, in habit.

Bull Springs, altitude 1,900 m., July 26 (No. 1036).

525

Allium reticulatum Don, Mem. Wern. Soc. vi, 36 (1826–1831).
Edgemont, altitude 1,050 m., June 13 (No. 1037).

Leucocrinum montanum Nutt.; Gray. Ann. Lyc. N. Y. iv, 110 (1837).
Common around Custer, altitude 1,600 to 1,700 m., May 30 (No. 1038).

Polygonatum biflorum commutatum (Roem. & Schult.) Morong. Bull. Torr. Club, xx, 480 (1893); *Convallaria commutata* Roem. & Schult. Syst. vii, 1671 (1830).
Lead City, altitude 1,600 m., July 6 (No. 1039).

Vagnera amplexicaulis (Nutt.) Greene, Man. Bay Reg. 316 (1894); *Smilacina amplexicaulis* Nutt. Journ. Acad. Phila. vii, 58 (1831).
Wooded hillside: Lead City, altitude 1,600 m., July 6 (No. 1040).

Vagnera stellata (L.) Morong, Mem. Torr. Club. v, 114 (1894); *Convallaria stellata* L. Sp. Pl. i, 316 (1753).
Rare: Elk Canyon, altitude 1,200 m., June 29 (No. 1042).
A form with narrower, conduplicate leaves and slightly longer pedicels, corresponding to *Unifolium liliaceum* Greene, was also found. Hot Springs, altitude 1,100 m., June 11 (No. 1041).

Unifolium canadense (Desf.) Greene, Bull. Torr. Club, xv, 287 (1888); *Maianthemum canadensis* Desf. Ann. Mus. Par. ix, 54 (1807).
In shady woods: Little Elk Canyon, altitude 1,100 m., June 27 (No. 1043).

Yucca glauca Nutt. Fraser's Cat. 1813.
Among the foothills, east of Custer, altitude 1,400 m., July 23 (No. 1044).

Lilium umbellatum Pursh, Fl. i, 229 (1814).
In woods: Little Elk Canyon, altitude 1,200 m., June 27 (No. 1045).

Calochortus gunnisoni Wats. Bot. King Surv. 348 (1871).
In all specimens collected by me, the bulb was only a few inches below the surface, and at the time of blooming without a secondary bulb. Hill above Whitewood, altitude 1,200 m., July 7 (No. 1046).

Calochortus nuttallii Torr. & Gr. Pac. R. Rep. ii, pt. 2, 124 (1855).
In this the bulb was deep down, 15 to 25 cm. below the surface, at the time of blooming, often with a secondary small bulb a few cm. above the principal ones. Fall River Falls, altitude 1,000 m., June 19 (No. 1047).

Streptopus amplexifolius (L.) DC. Fl. Fran. iii, 174 (1805); *Uvularia amplexifolia* L. Sp. Pl. i, 304 (1753).
Near water: Sylvan Lake, altitude 1,900 m., July 20 and August 7 (No. 1048).

Disporum trachycarpum (Wats.) Benth. & Hook. Gen. Pl. iii, 832 (1883); *Prosartes trachycarpa* Wats. Bot. King Surv. 344 (1871).
In shady places below Sylvan Lake, altitude 1,900 m., June 8 and August 17 (No. 1019).

Zygadenus elegans Pursh, Fl. i, 241 (1814).
Little Elk River, altitude 1,200 m., June 27; Elk Canyon, altitude 1,300 m.,. June 29; Bull Springs, altitude 1,900 m., July 26 (No. 1050).

Zygadenus venenosus Wats. Proc. Amer. Acad. xiv, 279 (1879).
This species has been confused with *Z. nuttallii*. Hot Springs, altitude 1,100 m.,. June 13 (No. 1051).

SMILACACEÆ.

Smilax herbacea L. Sp. Pl. ii, 1030 (1753).
Hot Springs, altitude 1,050 m., June 19 (No. 1052).

COMMELINACEÆ.

Tradescantia virginiana L. Sp. Pl. i, 288 (1753).
Hot Springs, altitude 1,050 m., June 13 (No. 1053).

JUNCACEÆ.

Juncus vaseyi Engelm. Trans. St. Louis Acad. ii, 418 (1866).
In habit very much resembling the next, from which it differs in the longer, narrower capsule and the tailed seeds. It was growing together with the next at Hot Springs, altitude 1,050 m., August 3 (No. 1051).

Juncus tenuis Willd. Sp. Pl. ii, 214 (1799).
Hot Springs, altitude 1,050 m., August 3; Custer, altitude 1,650 m., May 30 (No. 1055).

Juncus bufonius L. Sp. Pl. i, 328 (1753).
Rare: Hermosa, altitude 1,050 m., June 22 (No. 1056).

Juncus longistylis Torr. Bot. Mex. Bound. 223 (1859).
Not common: Lead City, altitude 1,700 m., July 6 (No. 1057).

Juncus xiphioides montanus Engelm. Trans. Acad. St. Louis, ii, 481 (1868).
Rare; wet meadow: Custer, altitude 1,600 m., August 16 (No. 1058).

Juncus nodosus L. Sp. Pl. ed. 2, i, 466 (1762).
Banks of Fall River, near Hot Springs, altitude 1,050 m., August 3 (No. 1060).

Juncus torreyi Coville, Bull. Torr. Club, xxii, 303 (1895); *Juncus nodosus megacephalus* Torr. Fl. N. Y. ii, 326 (1843), not *J. megacephalus* Curtis.
With the preceding (No. 1061).

Juncoides comosum (Meyer) Sheldon, Bull. Geol. Nat. Hist. Surv. Minn. ix, 61 (1894); *Luzula comosa* Meyer, Syn. Luz. 18 (1823).
Rare: Elk Canyon, altitude 1,200 m., June 29 (No. 1062).

ALISMACEÆ.

Sagittaria arifolia (Nutt.) J. G. Smith, Ann. Rep. Mo. Bot. Gard. vi. [reprint 6] (1894).
The akenes are not mature enough for satisfactory identification. Custer, altitude 1,600 m., July 16; Hot Springs, altitude 1,050 m., August 3 (No. 1063).

ZANNICHELLIACEÆ.

Potamogeton pectinatus L. Sp. Pl. i, 127 (1753).
In the warm springs above Hot Springs, altitude 1,050 m., August 3 (No. 1064).

Potamogeton foliosus Raf. Med. Rep. ser. 2, v, 354 (1808).
In the warm springs: Hot Springs, altitude 1,050 m., June 15 (No. 1065).

LEMNACEÆ.

Lemna minor L. Sp. Pl. ii, 970 (1753).
Rapid Creek, altitude 1,000 m., July 25 (No. 1066).

CYPERACEÆ.

Cyperus aristatus Rottb. Desc. & Icon. 23 (1773).
Rare: only a few small specimens collected in Ruby Glen, Custer, altitude 1,700 m., August 19 (No. 1067).

Cyperus acuminatus Torr. & Hook. Ann. Lyc. N. Y. iii, 435 (1836).
Wet meadow: Custer, altitude 1,700 m., July 16 (No. 1068).

Scirpus americanus Pers. Syn. i, 68 (1805).
Very rare; only one poor specimen secured: Elk Canyon, altitude 1,200 m., June 29 (No. 1069).

Scirpus lacustris L. Sp. Pl. i, 48 (1753).
Elk Canyon, altitude 1,200 m., June 29 (No. 1070).

Scirpus atrovirens pallidus Britton, Trans. N. Y. Acad. ix, 11 (1889).
In French Creek, at Custer, altitude 1,600 m., August 1 (No. 1071).

Scirpus cyperinus (L.) Kunth, Enum. ii, 170 (1837); *Eriophorum cyperinum* L. Sp. Pl. ed. 2, i, 77 (1762).
Rare: Custer, altitude 1,600 m., July 16 (No. 1072).

Scirpus pauciflorus Lightf. Fl. Scot. 1078 (1777).
Banks of French Creek, Custer, altitude 1,600 m., July 16 (No. 1073).

Eleocharis palustris (L.) Roem. & Schult. Syst. Veg. ii, 151 (1817); *Scirpus palustris* L. Sp. Pl. i, 47 (1753).

The specimens of the form most common in the Black Hills are slender and resemble variety *glaucescens* in habit, but the tubercle is rhomboidal, constricted below. Lead City, altitude 1,600 m., July 6; Custer, altitude 1,600 m., July 16 (No. 1074).
At Hot Springs, altitude 1,060 m., August 3, specimens were collected which had a taller, flattened culm, 6 to 8 dm. high; finely striate and purplish at the base; spikes large with thick, scarious-margined bracts. No akenes were seen (No. 1075).

Eleocharis acuminata (Muhl.) Nees, Linnæa, ix, 291 (1835); *Scirpus acuminatus* Muhl. Gram. 27 (1817).
Low, about 2 to 2.5 dm. high, slender, flat, resembling *E. tenuis*, but the akenes are those of *E. acuminata*, viz, bluntly triangular, finely muricate, yellowish, with the tubercle small, pyramidal. Hot Springs, altitude 1,050 m., June 6; Hermosa, altitude 1,000 m., June 24 (No. 1076).

Carex straminea crawei [1] Boott, Ill. 121 (1862).
Rare: Hot Springs, altitude 1,050 m., August 3 (No. 1077).

Carex filifolia Nutt. Gen. ii, 204 (1818).
This is regarded as very good for "winter pasture," and very likely has a nutritive value. On a dry table-land: Hot Springs, altitude 1,100 m., June 11 (No. 1078).

Carex pennsylvanica Lam. Encycl. iii, 388 (1789).
A western form with very long leaves (over 1.5 dm. long), was found in the open valleys near Custer, altitude 1,700 m., August 1 (No. 1079).
A low form with short leaves was common in early spring in the same valleys; May 5 (No. 1080).

Carex marcida Boott; Hook. Fl. Bor. Amer. ii, 212 (1839).
Two forms were collected, which Professor Bailey doubtfully refers to this species. They are both too young for identification. One, more tufted and lower, was found in the open valleys near Custer, altitude 1,700 m., June 5 (No. 1081). The other, taller and more simple, was growing in a similar place, June 4 (No. 1084).

Carex richardsonii R. Br.; Richards. App. Frankl. Journ. 751 (1823).
Common throughout the open valleys around Custer, altitude 1,700 m., June 6 (No. 1082).

Carex stenophylla Wahl. Kongl. Sven. Vet. Akad. Handl. ser. 2, xxiv, 142 (1803).
The specimens are too young for identification, but are referred, subject to question, to this species. Open valley around Custer, June 1 (No. 1083).

Carex stricta Lam. Encycl. iii, 387 (1789).
A form of this species, very slender, with long, soft leaves. In a damp, shaded place below Sylvan Lake, altitude 1,900 m., June 9 and July 18 (No. 1085).

Carex siccata Dewey, Amer. Journ. Sci. x, 278 (1826).
Rare: on the railroad embankment in Elk Canyon, altitute 1,200 m., June 29 (No. 1086).

Carex festiva Dewey, Amer. Journ. Sci. xxix, 246 (1835).
Very rare: near Rapid Creek, Rochford, altitude 1,600 m., July 12 (No. 1087).

Carex utriculata Boott, Hook. Fl. Bor. Amer. ii, 221 (1839).
Wet meadow: Custer, altitude 1,650 m., July 16 (No. 1088).

[1] The Carices of this collection were determined by Prof. L. H. Bailey.

Carex longirostris Torr.; Schwein. Ann. Lyc. N. Y. i. 71 (1824).
In a shady, wet place below Sylvan Lake, altitude 1,900 m., July 19 (No. 1089).

Carex nebraskensis Dewey. Amer. Journ. Sci. ser. 2, xviii. 102 (1854).
Meadow, near Custer, altitude 1,050 m., July 16 (No. 1090).

Carex retrorsa Schwein. Ann. Lyc. N. Y. i. 71 (1824).
Rare: in wet meadow, near Custer, altitude 1,650 m., July 16 (No. 1091).

Carex deweyana Schwein. Ann. Lyc. N. Y. i, 65 (1824).
Rare: together with last, July 16 (No. 1092).

Carex aurea Nutt. Gen. ii. 205 (1818).
Hills near Lead City, altitude 1,600 m., July 6 (No. 1093a).

Carex varia Muhl.; Wahl. Kongl. Sven. Vet. Akad. Handl. ser. 2, xxiv. 159 (1803).
Wet places in Elk Canyon, altitude 1,200 m., June 29 (No. 1094).

Carex laxiflora blanda (Dewey) Boott, Ill. 37 (1858); C. blanda Dewey. Amer. Journ. Sci. x. 45 (1826).
Sylvan Lake, July 18 (No. 1095).

Carex tribuloides bebbii (Olney) Bailey. Mem. Torr. Club. i, 55 (1889); C. bebbii Olney, Exsicc. fasc. 2, no. 12 (1871).
Together with C. straminea crawei at Hot Springs, altitude 1,650 m., August 3 (No. 1209a).

Carex tenella Schk. Riedgr. 23 (1801).
Rare: together with C. deweyana and C. retrorsa, below Sylvan Lake, altitude 1,900 m., July 18 (No. 1210a).

GRAMINEÆ.

Panicum capillare L. Sp. Pl. i, 58 (1753).
A very small and slender form, the same as No. 1788 of my collection from the sand Hills of central Nebraska. Hot Springs, altitude 1,050 m., August 9 (No. 1096).

Panicum virgatum L. Sp. Pl. i, 59 (1753).
Hillside, near Fall River Falls, altitude 1,000 m., July 10 (No. 1097).

Panicum scoparium Lam. Encycl. iv. 744 (1797).
Hill, Lead City, altitude 1,700 m., July 9 (No. 1098).

Panicum dichotomum L. Sp. Pl. i, 58 (1753).
A low and hairy form. Bull Springs in the Limestone District, altitude 1,900 m., July 27 (No. 1099).

Panicum depauperatum Muhl. Descr. Gram. 112 (1817).
Dry hills: Lead City, altitude 1,700 m., July 4; Custer, altitude 1,700 m., July 18 (No. 1100).

Panicum crus-galli L. Sp. Pl. i, 56 (1753).
A low and smooth form. Hot Springs, altitude 1,100 m., June 13 (No. 1101).

Setaria viridis (L.) Beauv. Agrost. (1812); Panicum viride L. Sp. Pl. ed. 2, i, 83 (1762).
In the specimens collected, the bristles are unusually long and generally purplish. Railroad embankment near Minnekahta, altitude 1,270 m., August 4 (No. 1102).

Spartina cynosuroides (L.) Willd. Enum. 80 (1809); Dactylis cynosuroides L. Sp. Pl. i, 71 (1753).
Custer, altitude 1650 m., July 16 (No. 1103).

Beckmannia crucæformis (L.) Host, Gram. Austr. iii. 5, t. 6 (1805); Phalaris crucaformis L. Sp. Pl. i. 55 (1753).
In a pond north of Custer, altitude 1,650 m., July 16 (No. 1093b).

Andropogon provincialis Lam. Encycl. i. 376 (1783).
A glaucous form approaching A. hallii. Minnekahta Plains, altitude 1,300 m., August 5 (No. 1104).

Andropogon scoparius Mx. Fl. i, 57 (1803).
A wholly smooth form, tufted, with flattened sheaths. Minnekahta Plains, altitude 1,300 m., August 5 (No. 1105).

Phalaris arundinacea L. Sp. Pl. i, 55 (1753).
In a stream near Buffalo Gap, altitude 975 m., June 21 (No. 1106).

Savastana odorata (L.) Scribner, Mem. Torr. Club, v, 31 (1894); *Holcus odoratus* L. Sp. Pl. ii, 1048 (1753).
Rare: Pringle, altitude 1,500 m., August 5 (No. 1107).

Alopecurus geniculatus fulvus (Smith) Scribn. Mem. Torr. Club, v, 38 (1894); *A. fulvus* Smith, Engl. Bot. t. 1467 (1793).
Common: Elk Canyon, altitude 1,200 m., June 29; Rochford, altitude 1,650 m., July 11; Custer, altitude 1,600 m., July 16 (No. 1108).

Phleum pratense L. Sp. Pl. i. 59 (1753).
Near a brook, south of Lead City, altitude 1,600 m., July 9 (No. 1109).

Stipa spartea Trin. Mem. Acad. St. Petersb. ser. 6, i, 82 (1829).
Hills: Custer, altitude 1,700 m., August 16 (No. 1110).

Stipa comata Trin. & Rupr. Mem. Acad. St. Petersb. ser. 6, v, 75 (1842).
Hills: Custer, altitude 1,700 m., August 16 (No. 1111).

Stipa viridula Trin. Mem. Acad. St. Petersb. ser. 6, ii, 39 (1836).
Hills: Hot Springs, altitude 1,100 m., August 3 (No. 1112).

Stipa richardsonii Link, Hort. Berol. ii, 245 (1833).
This is the true *S. richardsonii* Link, according to Prof. F. Lamson-Scribner, not the plant so named in Gray's Manual, which is a distinct species, *S. macounii* Scribner. As most descriptions refer to this latter, I at first thought that my plant was a new species and described it as follows: Culms tufted from a short rootstock, slender, 6 to 9 dm. high, smooth; root leaves 1.5 to 2.5 dm. long, stiff, involute, from a loose sheath, minutely scabrous; panicle of slender, flexuose capillary branches, 1 to 1.5 dm. long, which are generally in pairs; outer glumes ovate, membranaceous above, hyaline and acute, unequal, both 3-nerved, purplish when young; flowering glumes only 4 mm. long, black when mature, thinly hairy all over; awn 15 to 25 mm. long, bent at the middle, the lower half twisted, slightly hairy. It much resembles *S. arcuacea*, but has a grain of only two-thirds the size and an awn scarcely one-half as long.
On wooded hills: Rochford, altitude 1,700 m., July 12; Custer, altitude 1,600 m., August 19 (No. 1113).

Oryzopsis asperifolia Mx. Fl. i. 51 (1803).
Both this and the next are wanting in Coulter's Manual. Sylvan Lake, altitude 1,800 m., June 8 (No. 1114).

Oryzopsis juncea (Mx.) B. S. P. Prel. Cat. N. Y. 67 (1888); *Stipa juncea* Mx. Fl. i, 54 (1803).
Together with the preceding (No. 1115).

Oryzopsis micrantha (Trin. & Rupr.) Thurb Proc. Acad. Phila. 1863, 78 (1863); *Urachne micrantha* Trin. & Rupr. Mem. Acad. St. Petersb. ser. 6, v, 16 (1842).
Rare: Elk Canyon, altitude 1,200 m., June 29 (No. 1116).

Oryzopsis cuspidata (Nutt.) Benth; Vasey, Grasses U. S. 23 (1883); *Eriocoma cuspidata* Nutt. Gen. i, 40 (1818); *Oryzopsis membranacea* (Pursh) Vasey, Grasses S. W. pt. 2, t. 10 (1891); *Stipa membranacea* Pursh, Fl. ii, 728 (1814), not L.
In canyons: Hot Springs, altitude 1,100 m., June 13 (No. 1117).

Aristida fasciculata Torr. Ann. Lyc. N. Y. i. 151 (1824).
Custer, altitude 1,700 m., August 16; Hot Springs, altitude 1,100 m., August 3 (No. 1118).

Muhlenbergia racemosa (Mx.) B. S. P. Prel. Cat. N. Y. 67 (1888); *Agrostis racemosa* Mx. Fl. i, 53 (1803).

A tall and leafy form which may perhaps be referred to the variety *ramosa*. Common on French Creek, altitude 1,600 m., July 22 (No. 1120).

A few specimens were collected near Custer, altitude 1,100 m., August 1, which differ in being more slender and in the empty glumes having longer awns (No. 1121).

Sporobolus cryptandrus (Torr.) Gray, Man. 576 (1848); *Agrostis cryptandra* Torr. Ann. Lyc. N. Y. i, 151 (1824).

Table-lands: Hot Springs, altitude 1,100 m., August 10 (No. 1122).

Sporobolus heterolepis Gray, Man. ed. i, 576 (1848); *Vilfa heterolepis* Gray. Ann. Lyc. N. Y. iii, 233 (1835).

Rare: Pringle, altitude 1,500 m., August 5 (No. 1123).

Sporobolus cuspidatus (Torr.) Scribner, Bull. Torr. Club x, 63 (1882); *Vilfa cuspidata* Torr.; Hook. Fl. Bor. Amer. ii, 238 (1840).

In Gray's list, Newton & Jenney's Report.[1]

Agrostis alba L. Sp. Pl. i, 63 (1753).

In wet meadows below Custer, altitude 1,600 m., August 1 (No. 1124).

Agrostis exarata Trin. Unill. 207 (1821).

Rare: Hot Springs, altitude 1,050 m., August 10 (No. 1125).

Agrostis hiemalis (Walt.) B. S. P. Cat. Pl. N. Y. 68 (1888); *Cornucopiæ hyemalis* Walt. Fl. Car. 74 (1788), teste Mx.

The specimens have broad, upright leaves. Wet meadow: Custer, altitude 1,600 m., July 16 (No. 1126).

Calamagrostis canadensis Mx. Beauv. Agrost. 15 (1812); *Arundo canadensis* Mx. Fl. i, 73 (1803).

In my specimens the leaves are more or less involute. Along Fall River, Hot Springs, altitude 1,050 m., August 10 (1128).

Calamagrostis canadensis dubia (Scribner) Vasey, Contr. Nat. Herb. iii, 80 (1892); *Deyeuxia dubia* Scribner, Bot. Gaz. xi, 171 (1886).

Wet meadow, below Custer, altitude 1,600 m., August 16 (No. 1127).

Calamagrostis neglecta (Ehrh.) Gaertn. Fl. Wett. i, 94 (1799); *Arundo neglecta* Ehrh. Beitr. vi, 137 (1791).

Rare: Hot Springs, altitude 1,050 m., August 3 (No. 1129).

Calamagrostis sylvatica americana Vasey, Contr. Nat. Herb. iii, 83 (1892).

Very probably this is a distinct species. It differs much from *C. sylvatica* of Europe. Woods: Rochford, altitude 1,650 m., July 11 (No. 1130).

Calamovilfa longifolia (Hook.) Hack. True Grasses. 113 (1890); *Calamagrostis longifolia* Hook. Fl. Bor. Amer. ii, 241 (1840).

Hot Springs, altitude 1,100 m., August 9 (No. 1131).

Avena striata Mx. Fl. i. 73 (1803).

In woods: Elk Canyon, altitude 1,200 m., June 30; Custer, altitude 1,700 m., July 16 (No. 1132).

Danthonia spicata (L.) Beauv.; Roem. & Schult. Syst. Veg. ii, 690 (1817); *Avena spicata* L. Sp. Pl. i, 80 (1753).

In woods, not uncommon: Rochford, altitude 1,700 m., July 12; Custer, altitude 1,650 m., August 16 (No. 1133).

Schedonnardus paniculatus (Nutt.) Trelease; Branner & Coville, Rep. Geol. Surv. Ark. 1888, pl. 1, 236 (1891); *Lepturus paniculatus* Nutt. Gen. i, 81 (1818).

Very rare: Hot Springs, altitude 1,100 m., June 19 (No. 1134).

Bouteloua hirsuta Lag. Var. Ciene. y Litter. ii. pl. 4, 111 (1805).
Prairies: Hot Springs, altitude 1,100 m., June 19 (No. 1135).

Bouteloua oligostachya (Nutt.) Torr.; Gray, Man. ed. 2, 553 (1856); *Atheropogon oligostachyus* Nutt. Gen. i, 78 (1818).
Prairies: Hot Springs, altitude 1,100 m.. June 19 (No. 1136).

Bouteloua curtipendula (Mx.) Torr. in Emory, Mil. Recon. 153 (1848); *Chloris curtipendula*, Mx. Fl. i, 59 (1803).
Rare: Hot Springs, altitude 1,100 m.. June 19 (No. 1137).

Bulbilis dactyloides (Nutt.) Raf.; Kuntze, Rev. Gen. Pl. ii, 763 (1891); *Sesleria dactyloides* Nutt. Gen. i, 65 (1818).
Prairies: Hot Springs, altitude 1,100 m., June 13; on the French Creek, east of Custer, altitude 1,100 m., July 18 (No. 1138).

Kœleria cristata (L.) Pers. Syn. i, 97 (1805); *Aira cristata* L. Sp. Pl. i, 63 (1753).
Common: Hot Springs, altitude 1,100 m.. June 18; Elk Canyon, altitude 1,200 m., June 30; Lead City, altitude 1,700 m., July 9; Rochford, altitude 1,700 m., July 12 (No. 1139).

Eatonia pennsylvanica (DC.) Gray, Man. ed. 2, 558 (1856); *Kœleria pennsylvanica* DC. Cat. Hort. Monsp. 117 (1813).
Rare: Hot Springs, altitude 1,050 m., June 15 (No. 1140).

Catabrosa aquatica (L.) Beauv. Agrost. 157 (1812); *Aira aquatica* L. Sp. Pl. i, 61 (1753).
In a swamp near Pringle, altitude 1,500 m., August 5 (No. 1141).

Eragrostis major Host. Gram. Austr. iv, 14 (1809).
Rare: Hot Springs, altitude 1,050 m., August 9 (No. 1142).

Dactylis glomerata L. Sp. Pl. i, 71 (1753).
Rare: Hot Springs, altitude 1,050 m., June 15 (No. 1143).

Poa fendleriana (Steud.) Vasey, Ill. N. A. Grasses, ii, 71 (1893); *Eragrostis fendleriana* Steudel, Syn. Pl. Gram. 278 (1855).
The panicle is more open than usual, and the glumes are very light in color and shining. It was growing in big tufts on the prairies south of Pringle, altitude 1,500 m., August 5 (No. 1144).
A few bunches with broader, flat leaves and greener flowers were found at Hot Springs, altitude 1,100 m., June 13 (No. 1145).

Poa tenuifolia Buckley, Proc. Acad. Phila. 1862, 96 (1862).
Of two forms collected, one is tall, 1 to 5 dm. high, with broader leaves. Elk Canyon, altitude 1,200 m., June 29; Hot Springs, altitude 1,100 m., August 3 (No. 1146). Another form, referred to this species by Professor Scribner, is densely tufted, 1 to 2 dm. high, scabrous; leaves 3 to 7 cm. long, very narrow, soon involute, scabrous; panicle 5 to 7 cm. long, narrow, with short, upright branches. It differs from the typical form in size, in the narrow, scabrous leaves, the smaller and more rounded spikelets, and the broader glumes. Dry soil: Hot Springs, altitude 1,100 m., June 13 (No. 1147).

Poa nevadensis Vasey, Bull. Torr. Club, x, 66 (1883).
On the railroad embankment above Custer, altitude 1,650 m., July 16 (No. 1148).
Together with the more typical form, another was growing that had a very thick and dense panicle 10 to 15 cm. long and over 2 cm. wide, and large 5- to 8- flowered spikelets, about 1 cm. long, on a short pedicel (No. 1149).

Poa annua L. Sp. Pl. i, 68 (1753).
Rare: Elk Canyon, altitude 1,200 m., June 29 (No. 1150).

Poa pseudopratensis[1] Scribner & Rydberg, sp. nov.   Pl. XX.
Culms erect, 1 to 2 feet high from a creeping rootstock. Sheaths smooth or very

[1] The description of this species is drawn by Prof. F. Lamson-Scribner.

minutely scabrous; ligule scarious, acute, about 2 lines long, decurrent; leaf blade flat, 1 to 3 lines wide, those of the culm 1 to 3 inches long, those of the sterile shoots 6 to 10 inches long, midnerve prominent beneath, smooth on both surfaces except near the rigid acute tips, the distinctly cartilaginous margins scabrous. Panicle 2 to 4 or 5 (usually about 3) inches long, the scabrous branches at first nearly erect, widely spreading in anthesis; spikelets 3- to 5-flowered, 3 to 4 lines long, usually much longer than the rough pedicels; empty glumes nearly equal, 3-nerved, broadly lanceolate, acute with scarious margins and tips, the keel of the larger second glume scabrous near the apex; flowering glumes oblong, obtuse, 5-nerved, with scarious margins, silky-hairy on the nerves to near the middle and pubescent all over on the dorsal surface near the base, minutely scabrous in the upper part; palea as long as the glume, ciliate-scabrous on the keels, villous near the base.

It has been doubtfully referred to *P. pratensis*. From this it differs in its longer and *acute* ligule, its larger spikelets, and its less strongly compressed glumes, which have broader scarious margins and no cobweb at the base. It resembles also, somewhat, *P. alpina*, but differs in its larger size, long, creeping rootstock, long, *acute* ligule, and empty glumes not conspicuously crested on the keel.

Hot Springs, altitude 1,050 m., June 13; Custer, altitude 1,650 m., July 16 (No. 1151). No. 1272 from the Sand Hills of central Nebraska is the same. It has also been collected by John Macoun at Cypress Hills, British America, in August, 1880, and by Mrs. S. B. Walker at Castle Rock, Colo., in 1890.

Poa pratensis L. Sp. Pl. i, 67 (1753).

A variable species, the extreme forms of which seem very different from each other. One form, very low and tufted with very narrow leaves and small spikelets, was found near Lead City, altitude 1,600 m., July 6 (No. 1153). A form 5 to 8 dm. high, with broad and long leaves (15 to 20 cm. long and 6 mm. wide), and very large spikelets was collected at Hot Springs, altitude 1,050 m., June 13 (No. 1156). A similar one, but with narrow panicle as in *P. serotina*, was found in Elk Canyon, altitude 1,200 m., June 30 (No. 1157). These are perhaps distinct from *P. pratensis*.

Poa nemoralis L. Sp. Pl. i. 69 (1753).

Several forms were collected, which have all been referred to this species by Professor Scribner. One is a low plant approaching variety *stricta*, but having a more open panicle. It resembles *P. cæsia* collected by Rusby in Arizona. Hot Springs, altitude 1,050 m., June 11 (No. 1155). A form that by several has been mistaken for *P. serotina* was collected at Lead City, altitude 1,600 m., July 6: Custer, altitude 1,700 m., August 1 (No. 1158). It has still leaves and spreading panicle. *P. serotina* of C. C. Parry's collection and perhaps of Watson, King Survey collection, is the same. Another form is like the last, but with smaller and light-colored spikelets and broader, more flaccid leaves. In a wet place: Rapid City, altitude 1,000 m., June 25 (No. 1159). A slender form with spreading panicle, small, light-green spikelets, and longer pedicels was mistaken for *P. alsodes*. Lead City, altitude 1,600 m., July 6 (No. 1160).

Panicularia nervata (Willd.) Kuntze, Rev. Gen. Pl. ii, 783 (1891); *Poa nervata* Willd. Sp. Pl. i, 389 (1798).

In wet meadows: Whitewood, altitude 1,100 m., July 7 (No. 1161).

Panicularia americana (Torr.) MacMillan, Metasp. Minn. Val. 81 (1892); *Poa aquatica americana* Torr. Fl. U. S. i, 108 (1824).

In ponds, above Custer, altitude 1,650 m., July 16 (No. 1163).

Festuca ovina L. Sp. Pl. i, 73 (1753) var.

The specimens are low, with a narrow panicle, and short, narrow leaves. It resembles the variety *pseudo-ovina* Hack. It grows in bunches on dry prairie. Lead City, altitude 1,600 m., July 4 (No. 1161).

PLATE XX.

POA PSEUDOPRATENSIS Scribner & Rydberg.

Bromus kalmii Gray, Man. ed. i. 600 (1848).

Lead City, altitude 1,600 m., July 4; Rochford, altitude 1,700 m., July 12; Hot Springs, altitude 1,100 m., August 8 (No. 1165).

Bromus ciliatus L. Sp. Pl. i, 76 (1753).

Rare: Rochford, altitude 1,700 m., July 12 (No. 1166).

Bromus pumpellianús Scribner, Bull. Torr. Club, xv, 9 (1888).

Hillsides: Runkels, altitude 1,300 m., June 30; Rochford, altitude 1,700 m., July 12; Custer, altitude 1. 700 m., July 16 (No. 1167).

Agropyron repens glaucum (Desf.) Scribner, Mem. Torr. Club, v. 57 (1891); *Triticum glaucum* Desf. Tabl. Bot. Mus. 16 (1801).

Hills below Deadwood, altitude 1,500 m., July 5 (No. 1168).

Agropyron violaceum majus Vasey, Contr. Nat. Herb. i, 280 (1893).

These specimens seem to belong to this variety, which, however, I do not think is a variety of *A. violaceum*. Professor Scribner regards them as a form of *A. repens*, which they resemble very much. I should take them for a form of that species if it were not for the fact that I could not find any creeping rootstock. They were growing in clumps in the manner of *A. tenerum*. Deadwood, altitude 1,500 m., July 5 (No. 1170). A form of the same, with awns over 1 cm. long, was collected at Rochford, altitude 1,700 m., July 12 (No. 1171).

Agropyron tenerum Vasey, Bot. Gaz. x, 258 (1885).

Deadwood, altitude 1,500 m., July 5 (No. 1169).

Agropyron caninum (L.) Roem. & Schult. Syst. Veg. ii, 756 (1817); *Triticum caninum* L. Sp. Pl. i, 86 (1753).

Common: Lead City, altitude 1,600 m., July 6; Custer, altitude 1,700 m., July 16 and August 16; Hot Springs, altitude 1,100 m., August 3 (No. 1172).

Elymus canadensis L. Sp. Pl. i, 83 (1753).

Along Fall River: Hot Springs, altitude 1,050 m., August 9 (No. 1173).

Elymus canadensis glaucifolius (Willd.) Torr. Fl. U. S. i, 137 (1821); *Elymus glaucifolius* Willd. Enum. 131 (1809).

Hot Springs, altitude 1,050 m., August 9 (No. 1174).

Elymus virginicus L. Sp. Pl. i, 84 (1753), var.

This is the same as No. 1553 of my collection from the Sand Hills of central Nebraska. Hot Springs, August 9 (No. 1175).

Elymus striatus Willd. Sp. Pl. i, 470 (1797).

Rare, with the preceding three. Hot Springs, August 9 (No. 1176).

Elymus elymoides (Raf.) Swezey, Cat. Nebr. Pl. 15 (1891); *Sitanion elymoides* Raf. Journ. Phys. lxxxix, 103 (1819).

Rare, on dry prairie: Hot Springs, altitude 1,100 m., June 13 and August 9 (No. 1177.)

Elymus dasystachys Trin.; Ledeb. Fl. Alt. i, 120 (1831).

The specimens in the collection have much shorter spikes and larger and more hairy spikelets than in the Siberian form. Elk Canyon, near Runkels, altitude 1,300 m., June 30 (No. 1178).

Hordeum jubatum L. Sp. Pl. i, 85 (1753).

Custer, altitude 1,650 m., July 16 (No. 1179).

CONIFERÆ.

Juniperus communis sibirica (Burgsd.); *Juniperus sibirica* Burgsd. Anl. Erz. Anpfl. Holzart. ii, 272 (1787); *J. communis alpina* Gaud. Fl. Helv. vi, 301 (1830).

The name of this plant has been changed lately by botanists in this country to *J. nana* Willd.[1] Willdenow cites *J. sibirica* Burgsd. as a synonym. I have not been

---

[1] Sp. Pl. iv, 851 (1806).

able to see Burgsdorf's original description, but the variety was already known and
had been described by Linnæus in the Species Plantarum, and by Pallas in the Flora
Rossica, although not named. Taking Willdenow as authority, I adopt the name
sibirica. Should Willdenow have been mistaken, and Burgsdorf's shrub not have
been the same as his (the identity has not been denied), there is one more name older
than Willdenow's that has to be taken into consideration, viz, *J. communis montana*
Ait.,[1] the identity of which with Linnæus variety is not questionable. As to the
relationship to *J. communis*, I am of the opinion that this plant is best considered as
a variety of that species, as many intermediate forms are found.

    Custer, altitude 1,800 m., June 4 (No. 1180).

Juniperus sabina prostrata (Pers.) Loud. Arbor. Frut. Brit. iv, 2489 (1838); *J. prostrata* Pers. Syn. Pl. ii, 632 (1807).

    This name should be used instead of *J. sabina procumbens* Pursh, unless *J. horizontalis* Moench,[2] is the same. I have no means of verifying the identity of the two.
The American trailing savin is well distinguished from *J. sabina* of Europe. Koch
and Gordon regard it as a distinct species, and to merge it in *J. sabina*, as has been
done lately, is unwarranted. On dry foothills: Hermosa, altitude 1,100 m., June 23;
Piedmont, altitude 1,100 m., June 27 (No. 1181).

Juniperus virginiana L. Sp. Pl. ii, 1039 (1753).

    Very rare in the hills proper; only two shrubs seen, on the Buckhorn Mountain,
near Custer, altitude 1,800 m., June 4. More common in the foothills: Hot Springs,
altitude 1,100 m., June 15. One shrub at the latter place had both male and female
flowers (No. 1182).

Pinus ponderosa scopulorum Engelm. Bot. Cal. ii, 126 (1880).

    Common throughout the Black Hills. The Northern Hills were formerly covered
with forest, in which this was the predominant species, but a large portion of the
tract has been devastated by mining companies and sawmill operators. Hot
Springs, altitude 1,100 m., June 17 (No. 1183).

Picea canadensis (Mill.) B. S. P. Cat. Pl. N. Y. 71 (1888): *Abies canadensis* Mill.
Gard. Dict. ed. 8, no. 4 (1768).

    Not uncommon in the higher hills, especially on the northern sides. Fruit: Rochford, altitude 1,600 m., July 12 (No. 1210 b).

SELAGINELLACEÆ.

Selaginella rupestris (L.) Spring, in Mart. Fl. Bras. i, pt. 2, 118 (1840); *Lycopodium rupestre* L. Sp. Pl. ii 1101 (1753).

    On dry hills: local: Custer, altitude 1,700 m., June 6 (No. 1181).

LYCOPODIACEÆ.

Lycopodium obscurum L. Sp. Pl. ii, 1102 (1753); *L. dendroideum* Mx. Fl. ii, 282 (1803).

    Michaux's species seems to be the true *L. obscurum* L.

    Elk Canyon, altitude 1,200 m., June 29 (No. 1185).

OPHIOGLOSSACEÆ.

Botrychium matricariæfolium (?) A. Br. in Doell, Rhein. Fl. 24 (1843).

    It seems to stand nearest this species, but the sterile frond is sessile. I took it to
be a form of *B. boreale* Milde, with the description of which it agrees quite well. It
differs, however, from European specimens in the National Herbarium, in the more
slender habit, and in the smaller and less crowded divisions of the sterile frond,

---

[1] Hort. Kew. iii, 414 (1789).

[2] Meth. Pl. Hort. and Ag. Marburg, 699 (1794).

which is ovate-oblong in outline, not broadly triangular-ovate. Only two specimens (15 to 20 cm. high) were collected by me, on a shaded hillside south of Custer, altitude 1,700 m., August 15. A few specimens were also collected by Prof. A. F. Woods and one of the students of the University of Nebraska. The specimens are of a form that seems to be intermediate between *B. lunaria*, *B. boreale*, *B. lanceolatum*, and *B. matricariæfolium*. It may be a new species, but the material is too meager to warrant a publication (No. 1186).

POLYPODIACEÆ.

Polypodium vulgare L. Sp. Pl. ii. 1085 (1753).
Common in crevices of rocks around Custer, altitude 1,700 m., July 18 (No. 1187).

Polypodium vulgare rotundatum Milde, Fil. Eur. & Atlan. 18 (1867).
It differs from the preceding in its short fronds with rounded lobes and its larger, more confluent sori. Wheeler's Expedition, No. 992, and Watson's No. 1357, belong also to this variety, which has not hitherto been reported for America. In crevices: Custer, altitude, 1,700 m., July 16 (No. 1188).

Cheilanthes gracilis (Fee) Mett. Abh. Senck. Nat. Gesell. iii [reprint 36] (1859); *Myriopteris gracilis* Fee, Gen. Fil. 150 (1850-1852).
On exposed rocks: Hot Springs, altitude, 1,100 m., June 11 (No. 1189).

Pellæa atropurpurea (L.) Link, Fil. Hort. Berol. 59 (1841); *Pteris atropurpurea* L. Sp. Pl. ii, 1076 (1753).
Canyon near Hot Springs, altitude 1,100 m., June 11 (No. 1190).

Pellæa breweri Eaton, Proc. Amer. Acad. vi, 555 (1865).
The specimens in this collection have fronds that are decidedly coriaceous, a modification probably due to the exposed locality in which they grew. I took them first to be a depauperate form of *P. atropurpurea*, but the divisions even of the fertile fronds are broadly ovate, the rachis bright brown instead of purplish black and without scales. They are brittle and when old show the depressions that make them look as if articulated, a characteristic of *P. breweri*. The fronds are 0.5 to 1 dm high from a tufted, thick rootstock, once pinnate, of 5 to 9 pinnæ; pinnæ, 1 cm. or more long, oval or ovate, entire, or the lower with a small lobe on the upper side.
In crevices of exposed limestone rocks, generally on the sunniest side: near Bull Springs, altitude 1,900 m., July 27 (No. 1191).

Pteris aquilina L. Sp. Pl. ii, 1075 (1753).
Custer, altitude 1,700 m., August 19 (No. 1192).

Asplenium trichomanes L. Sp. Pl. ii, 1080 (1753).
Crevices of rocks below Sylvan Lake, altitude 1,900 m., August 18 (No. 1193).

Asplenium septentrionale (L.) Hoffm. Deutsch. Fl. ii, 12 (1795); *Acrostichum septentrionale* L. Sp. Pl. ii, 1068 (1753).
Crevices of rocks, especially on the north side of the mountains: Custer, altitude 1,700 m., June 5 and August 16 (No. 1194).

Asplenium filix-fœmina (L.) Bernh. Schrad. Neues Journ. Bot. i, pt. 2, 26 (1806); *Polypodium filix-fæmina* L. Sp. Pl. ii, 1090 (1753).
Common around Sylvan Lake, altitude 2,000 m., July 20 (No. 1195).

Phegopteris dryopteris (L.) Fee, Gen. Fil. 243 (1850-1852); *Polypodium dryopteris* L. Sp. Pl. ii, 1093 (1753).
In dark woods near Custer, altitude 1,700 m., August 19 (No. 1196).

Dryopteris filix-mas (L.) Schott, Gen. Fil. (1834); *Polypodium filix-mas* L. Sp. Pl. ii, 1091 (1753).
Among rocks: Rochford, altitude 1,200 m., July 12; Buckhorn Mountain, near Custer, altitude 1,800 m., July 16 (No. 1197).

Cystopteris fragilis (L.) Bernh. Schrad. Neues Journ. Bot. i. pt. 2, 27 (1806); *Polypodium fragile* L. Sp. Pl. ii, 1091 (1753).

Throughout the Black Hills: Little Elk, altitude 1,100 m., June 27; Lead City, altitude 1,600 m., July 6; Custer, altitude 1,700 m., August 15 (No. 1198).

Woodsia oregana Eaton, Can. Nat. ii, 90 (1865).

Common throughout the Black Hills: Hermosa, altitude 1,100 m., June 23; Elk Canyon, altitude 1,200 m., June 29; Custer, altitude 1.700 m., August 10 (No. 1199).

Woodsia scopulina Eaton, Can. Nat. ii, 90 (1865).

On wooded hillsides south of Custer, altitude 1,700 m., August 10 (No. 1200).

Onoclea sensibilis L. Sp. Pl. ii, 1062 (1753).

In Gray's list, Newton & Jenney's Report.[1] Also collected by Prof. J. A. Williams, near Rapid City.

Onoclea struthiopteris (L.) Hoffm. Deutsch. Fl. ii, 11 (1795); *Osmunda struthiopteris* L. Sp. Pl. 1066 (1753).

In Gray's list only.

EQUISETACEÆ.

Equisetum arvense L. Sp. Pl. ii, 1061 (1753).

Unusually robust specimens, in damp woods below Sylvan Lake, altitude 1,800 m., June 8 (No. 1201).

Equisetum sylvaticum L. Sp. Pl. ii. 1061 (1753).

With the preceding, June 8 (No. 1202).

Equisetum lævigatum A. Br.; Engelm. Amer. Journ. Sci. xlvi. 87 (1844).

The two forms collected in Nebraska were also found here. The more robust with sessile spike, No. 1260 of my Nebraska collection, was collected in Elk Canyon, altitude 1,200 m., June 29 (No. 1203). The other one, of the same form as No. 1283 of the Nebraska collection, was found at Hot Springs, altitude 1,050 m., August 3 (No. 1204).

---

[1] Geol. Surv. Black Hills, 537 (1880).

INDEX.

III.

www.ingramcontent.com/pod-product-compliance
Lightning Source LLC
Chambersburg PA
CBHW020258090426
42735CB00009B/1131